Shrews and Moles of British Columbia

Also from the Royal BC Museum

Rodents and Lagomorphs of British Columbia
by David W. Nagorsen

Bats of British Columbia
by Cori L. Lausen, David W. Nagorsen, R. Mark Brigham and Jared Hobbs

Carnivores of British Columbia
by David F. Hatler, David W. Nagorsen and Alison M. Beal

Nature Guide to the Victoria Region
edited by Ann Nightingale and Claudia Copley

Royal BC Museum Handbook

SHREWS AND MOLES

OF BRITISH COLUMBIA

Second Edition

DAVID W. NAGORSEN AND NICK PANTER

VICTORIA, CANADA

Shrews and Moles of British Columbia, second edition

Published by the Royal BC Museum, 675 Belleville Street, Victoria, British Columbia, V8W 9W2, Canada.

The Royal BC Museum is located on the traditional territories of the Lekwungen (Songhees and Xwsepsum Nations). We extend our appreciation for the opportunity to live and learn on this territory.

See p. 225 for credits and copyright information for specific elements in the book.

Library and Archives Canada Cataloguing in Publication
Title: Shrews and moles of British Columbia / David W. Nagorsen and Nick Panter.
Other titles: Opossums, shrews, and moles of British Columbia
Names: Nagorsen, David W., author. | Panter, Nick, author. | Royal British Columbia Museum, issuing body.
Description: Second edition. | Previously published under title: Opossums, shrews, and moles of British Columbia. | Includes bibliographical references and index.
Identifiers: Canadiana (print) 20230490433 | Canadiana (ebook) 20230490441 | ISBN 9781039900035 (softcover) | ISBN 9781039900042 (EPUB) | ISBN 9781039900059 (PDF)
Subjects: LCSH: Shrews—British Columbia. | LCSH: Moles (Animals)—British Columbia.
Classification: LCC QL737.S7 N34 2024 | DDC 599.33—dc23

10 9 8 7 6 5 4 3 2 1

Printed and bound in Canada by Friesens.

Contents

Preface

Small and cryptic, shrews and moles represent some of the most obscure and least familiar of any British Columbian mammals. They are rarely observed by the general public or naturalists, with human encounters typically limited to kills brought in by the family DOMESTIC CAT or a dead animal found by chance. Despite more focus on small mammals in biodiversity studies by provincial wildlife and forestry agencies, only a few of the province's shrews or moles have received any detailed study—a contrast with the effort directed at other small mammals. Much remains to be learned about their distribution, habitat requirements, breeding biology and taxonomy in the province. We hope this book will stimulate more attention for these much-neglected mammals.

Since the *Opossums, Shrews and Moles of British Columbia* handbook was published by the Royal BC Museum 28 years ago, five shrew species have been added to the province's fauna. Three were only recently found to occur in BC; two are the result of taxonomic splitting based on recent genetic research. With more than half the province's shrew and mole species of conservation concern, we felt there was a need for a revised handbook that includes these new species, updates the taxonomy nomenclature and summarizes new findings on the natural history of these animals.

This handbook has a new introduction and an updated General Biology section to incorporate new findings. It also uses many colour images throughout, including for the identification keys. Most of the change is in the individual species accounts. Images of various identification traits now enhance the species descriptions, and distribution maps include new occurrence records and shaded areas that depict the expected range. We have added a conservation section to describe each species's conservation status and threats. In addition to listing subspecies, we expanded the taxonomy section of each account to incorporate the findings of recent genetic studies. The reader may note that we omitted the NORTH AMERICAN OPOSSUM from this edition. No new information for the BC population is available to update its account in the 1996 handbook; moreover, as a marsupial mammal, it has no close relationship to shrews or moles.

How to Use This Book

Similar to previous Royal BC Museum mammal handbooks, the book is divided into three major sections—the introduction and General Biology; Identification Keys; and Species Accounts. The introduction and General Biology provide basic background information on the classification, evolution, anatomy and ecology of shrews and moles. It is intended for any reader curious for a better understanding of these poorly known mammals. Sections on conservation and methods for studying shrews and moles should be useful for wildlife biologists and students. Students or biologists surveying small mammals will need to use the identification keys to make reliable identifications. Summaries of identification traits are also presented in two appendixes. The species accounts focus on natural history details for the 14 shrew and 3 mole species found in BC. Each account contains illustrations, descriptions and a range map. Few photos are available of live shrews and moles; therefore, we used Michael Hames's drawings from the 1996 edition of this book, three shrew drawings by Donald Gunn and photographs of museum study skins to illustrate species. If your primary interest is in identifying a shrew or mole, start with the species accounts. Their range maps will help narrow down a list of candidate species for your location. Moles and a few shrew species may be identifiable from illustrations and descriptions in the accounts. However, most shrews cannot be identified from their external features.

For two species of shrews that could be found in northern BC but lack documented BC occurrences, we excluded them from the identification keys and the appendixes of identification traits. We summarize their diagnostic traits and habitat where they are likely to occur in a Hypothetical Species section.

Some 207 references cited in the book are listed in a references section. Scientific names for common names of plants and animals that appear in the book are given in an appendix. A glossary defines various technical terms used. It may be particularly helpful in understanding the traits used in the identification keys and species descriptions. A detailed index will assist the reader to locate any subject.

Introduction

Origins and Classification

Traditionally, shrews and moles were families placed in the mammalian order Insectivora, a large order with nine families. However, mammalogists long suspected that this group was a taxonomic wastebasket made up of species that share similar morphological features and lifestyles but have very different ancestries. Recent research applying molecular genetics has unravelled the complex phylogeny of this group. It places the former members of the Insectivora into four distantly related mammalian orders, with shrews and moles in the order Eulipotyphla. The four extant families in the Eulipotyphla are the Erinaceidae (hedgehogs and gymnures), widely distributed across Africa and Eurasia; the Solenodontidae (solenodons), which are endemic to islands of the Greater Antilles in the Caribbean; and the Soricidae (shrews) and Talpidae (moles), which are both represented in BC.

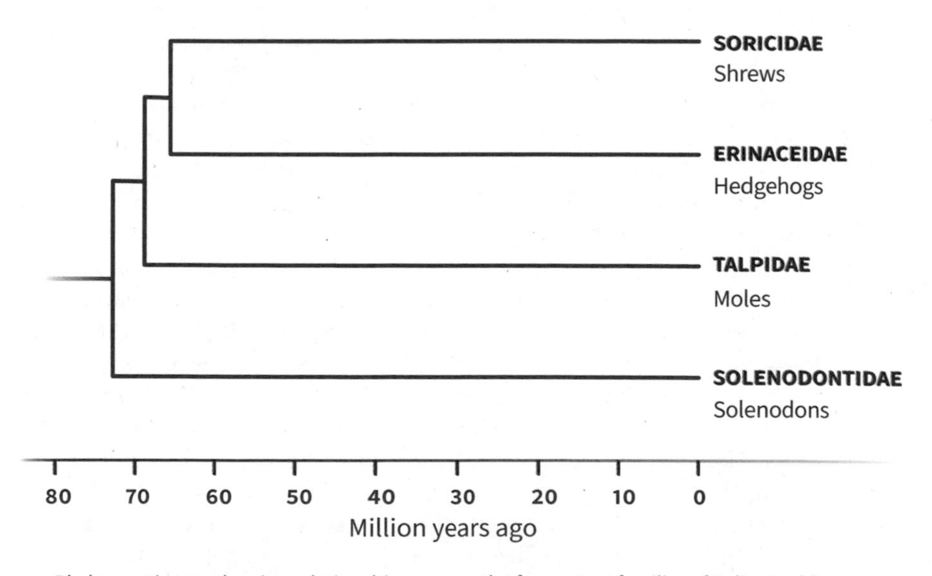

Phylogenetic tree showing relationships among the four extant families of Eulipotyphla.

The Eulipotyphla are an ancient lineage that originated some 74 million years ago in the late Cretaceous period. From 73 to 66 million years ago, this lineage split into the four families. Although moles and shrews were once considered to be the two most closely related families in the order, genetic studies clearly show that the nearest relatives to the shrews are the hedgehogs and gymnures. Taxonomists recognize about 557 species in the Eulipotyphla, with most (83 per cent) species belonging to the shrew family. After rodents and bats, this is the mammalian order with the third-greatest number of species.

Eulipotyphla literally translates as "truly blind and fat," but this hardly describes most animals in this group. Only a few moles are blind, and shrews can hardly be described as fat. The Eulipotyphla are members of the Laurasiatheria superorder, a major branch of the eutherian family tree, whose members include bats, carnivores and hoofed mammals.

Living species of Eulipotyphla have retained a number of primitive mammalian traits, and in many respects they resemble the ancestral stock that gave rise to the modern eutherian mammals. Among their primitive features are a generalized body plan with feet designed for walking on the soles, a relatively small brain in which the region associated with smell is the most developed, numerous sharp-cusped teeth that in many forms resemble the primitive dentition of eutherian mammals, and young that are born in a very immature, underdeveloped state. As is the case for all eutherian mammals, Eulipotyphla nourish their unborn young with a fully developed placenta. Despite their somewhat basic body plan, the Eulipotyphla are highly successful, and some demonstrate special adaptations for terrestrial, fossorial or semi-aquatic modes of life. The order includes some of the world's smallest mammals. The ETRUSCAN SHREW and HOLARCTIC LEAST SHREW, which weigh around two grams (the weight of a dime), are arguably the world's smallest mammals. A small body size has a profound influence on the life history of shrews and moles.

The Talpidae (moles) inhabit temperate regions of North America, Europe and Asia, with some 57 extant species recognized. There are seven species in North America, with three found in BC. According to molecular genetics data, the talpid lineage diverged from other eulipotyphlans in the late Cretaceous period. However, its geographical origins are unresolved. Until recently the oldest fossils (dating from the Eocene epoch) were from Europe or Asia, supporting the hypothesis that moles originated in Eurasia. However, the discovery of a fossil of similar age from the Florissant Fossil Bed Formation National Monument in Colorado raises the possibility of a North American origin for the talpids. Because these early fossils consist of isolated teeth or partial jaws with teeth, we know little about the actual lifestyle of these early moles and the extent to which they were adapted to a fossorial life. The living mole species demonstrate a remarkable diversity of life forms with an eclectic mix of most peculiar animals. Some mole species are fossorial forms adapted

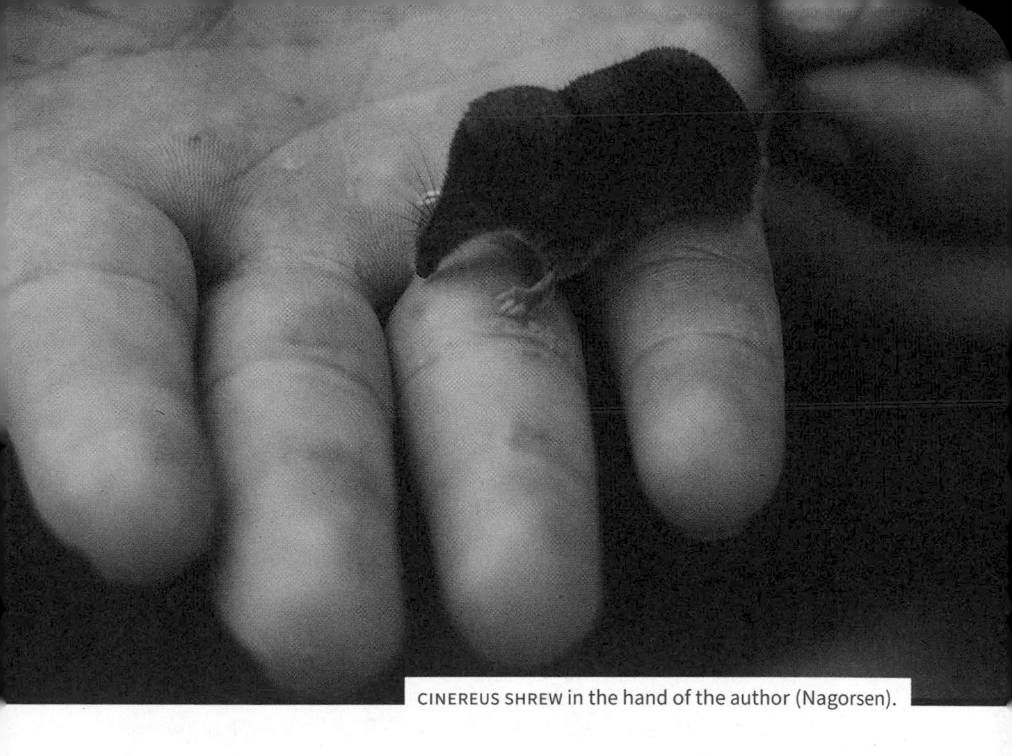
CINEREUS SHREW in the hand of the author (Nagorsen).

to a burrowing life underground, but the family also includes moles that are terrestrial (and shrew-like), semi-fossorial or highly aquatic. The three mole species found in BC are restricted to the Pacific coast region of North America. They include the semi-fossorial AMERICAN SHREW-MOLE and two truly fossorial moles: COAST MOLE and TOWNSEND'S MOLE.

The Soricidae (shrews) inhabit all of the major land areas except for Antarctica, Australia, New Zealand and southern South America with 459 extant species recognized. There are about 40 species in North America with 14 found in BC. Molecular genetic studies demonstrate that the Soricidae and Erinaceidae lineages diverged at the end of the Cretaceous period about 66 million years ago. The time coincides with a mass extinction event including the disappearance of the non-avian dinosaurs resulting from the earth's collision with a massive asteroid. One hypothesis is that the extinction of many vertebrates at the end of the Cretaceous opened up new opportunities for eulipotyphlan evolution. The Soricidae originated in Eurasia; the oldest known fossil, found in Germany, is from the Oligocene epoch. The family first appeared in North America in the middle Miocene epoch 13.9 to 12.1 million years ago.

Traditionally, the Soricidae are subdivided into two major groups: the red-toothed shrews (Soricinae) and white-toothed shrews (Crocidurinae), based on tooth colour and other morphological and physiological traits. Genetic studies confirmed that they are two distinct genetic lineages that separated in the early Miocene epoch some 18 million years ago. All North American shrews are members of the Soricinae. In contrast to the Talpidae with their diverse lifestyles, the Soricidae are rather generalized in their mode of locomotion, and most share a similar body form. A few have adapted to a semi-aquatic life. The most extreme example is the ELEGANT WATER SHREW of Asia, which has webbed front and hind feet and a keeled tail. The three species of water shrews found in BC are less specialized but have several adaptations for swimming and diving.

Taxonomy and Nomenclature

In 1753 the Swedish botanist Carl Linnaeus developed a classification system for all living organisms that is still used by biologists today. The Linnaean system is based on a hierarchy of taxonomic categories: class, order, family, genus and species. The species is the basic unit of classification, and species are grouped into higher taxonomic categories based on their presumed relationships. Each species has a unique scientific name (binomen) consisting of the genus name followed by the species name. By convention, the binomen is italicized, with the generic name capitalized and the specific name in lowercase. For example, the scientific name for the CINEREUS SHREW is *Sorex cinereus*. Closely related species that share a number of similar traits are usually grouped in the same genus. About 86 species are recognized in the shrew genus *Sorex*; all 14 shrews found in BC belong to this genus.

Traditionally, biologists have defined species as consisting of populations that are capable of interbreeding but are reproductively isolated from other populations. With the use of molecular genetics, recent species concepts downplay the strict criteria of a lack of interbreeding and instead focus on the use of genetic lineages to define species. This has resulted in several competing definitions and criteria, and the species concept continues to be a subject of debate. Some species have distinct geographical groups of populations that taxonomists recognize formally as subspecies with a trinomen. *Sorex pacificus prevostensis*, for example, is a large dark race of the PACIFIC SHREW restricted to Kunghit Island, in Haida Gwaii. It is one of the seven subspecies of PACIFIC SHREW found in the province. The subspecies concept has its defenders and critics; we have gone ahead and listed any recognized subspecies in our species accounts. Most subspecies listed for BC moles and shrews are based on morphological traits such as fur colour, and genetic studies are required to confirm the validity of these races.

Although the taxonomy of BC moles has been remarkably stable with no changes in species or subspecies names since the 1996 edition of this handbook, the recent application of molecular genetic techniques to sequence DNA and associated changes in how biologists define species have resulted in many recent changes in taxonomy and nomenclature affecting the province's shrews. A trend in taxonomy is splitting species into several geographically separated species that are genetically distinct but often demonstrate few morphological differences. The former Pygmy Shrew is now split into eastern and western species, and the coastal and Interior forms of the Dusky Shrew have been elevated to full species. The Water Shrew described in the original handbook is now split into three species, with two found in the province.

The scientific names for species and subspecies used in this book are based on a 2018 monograph by Neal Woodman that provides an exhaustive review of the taxonomy of living species of American Eulipotyphla. Any taxonomic changes since 2018 were taken from the Mammal Diversity Database (mammaldiversity.org), an online compendium of the world's mammals species that is periodically updated. There is no universally accepted list of mammalian English common names equivalent to the American Ornithological Society checklist of bird names. The Woodman publication gives no common names; therefore, we have based common names mostly on the Mammal Diversity Database. Be aware that common names for many mammals are somewhat contrived and are rarely used by scientists. Several shrews have alternate common names that are widely used.

List of BC Moles and Shrews

Families and genera are arranged according to the generally accepted phylogenetic order. Species within a genus are arranged alphabetically. Names of the authority who described the species follow the scientific name. If the species was originally described under a different genus, the authority is given in parentheses.

Order Eulipotyphla

Family Talpidae: Moles

Neurotrichus gibbsii (Baird)	AMERICAN SHREW-MOLE
Scapanus orarius True	COAST MOLE
Scapanus townsendii (Bachman)	TOWNSEND'S MOLE

Family Soricidae: Shrews

Sorex arcticus Kerr	ARCTIC SHREW
Sorex bendirii (Merriam)	PACIFIC WATER SHREW
Sorex cinereus Kerr	CINEREUS SHREW
Sorex eximius Osgood	WESTERN PYGMY SHREW
Sorex merriami Dobson	MERRIAM'S SHREW
Sorex navigator (Baird)	WESTERN WATER SHREW
Sorex obscurus Merriam	DUSKY SHREW
Sorex pacificus Coues	PACIFIC SHREW
Sorex palustris Richardson	AMERICAN WATER SHREW
Sorex preblei Jackson	PREBLE'S SHREW
Sorex rohweri Rausch, Feagin, Rausch	OLYMPIC SHREW
Sorex trowbridgii Baird	TROWBRIDGE'S SHREW
Sorex tundrensis Merriam	TUNDRA SHREW
Sorex vagrans Baird	VAGRANT SHREW

Family Soricidae: Hypothetical species
(no documented occurrences but may be found in BC)

Sorex haydeni Baird	PRAIRIE SHREW
Sorex minutissimus Zimmermann	HOLARCTIC LEAST SHREW

Shrews and Moles in BC

Encompassing some 950,000 square kilometres and spanning 11 degrees of latitude and 25 degrees of longitude, BC has the most diverse physiography and climate of any Canadian province. A series of north-south-oriented mountain ranges dominate the landscape. They play a major role in the province's climate by intercepting Pacific weather systems as they move eastward and creating alternating wet and dry belts. The wettest regions are along the Pacific coast, especially the western slopes of the Coastal mountain ranges and the outer coasts of Vancouver Island and Haida Gwaii. East of the Coast Mountains, rain shadows create a large arid region: the Interior Plateau. Most extreme arid conditions are found in some of the southern Interior valleys such as the Okanagan and Thompson River valleys. Other wet-dry belts are associated with the Cassiar, Rocky, Cariboo, Monashee, Selkirk and Purcell mountain ranges.

The vegetation of the province is predominantly coniferous forest, although deciduous forest is associated with northern boreal regions and riparian habitats along rivers and lakes. Grassland and shrub-steppe habitats occur in some of

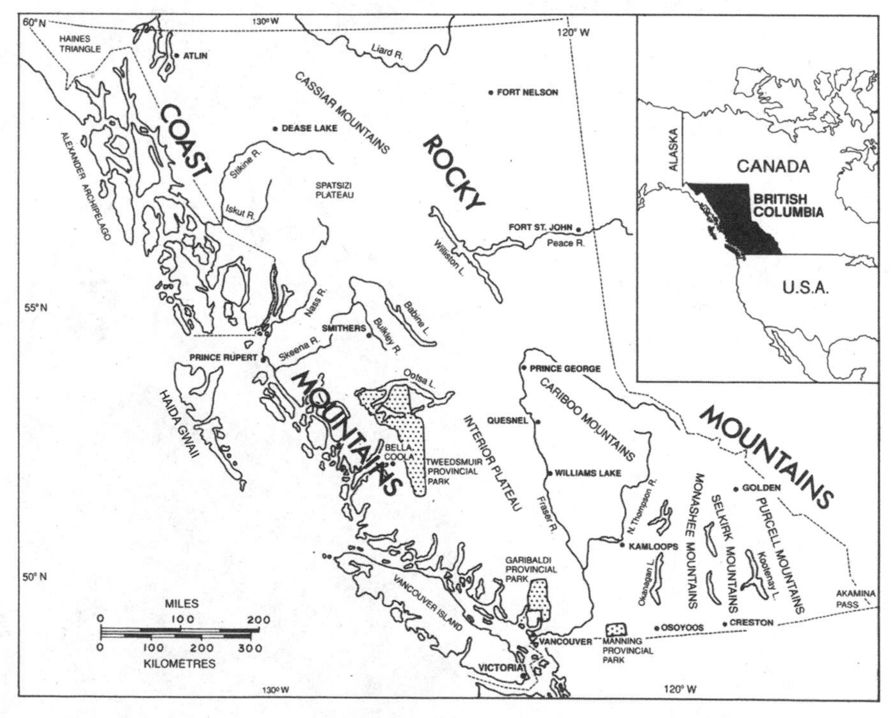

General geographical features of BC.

Above: Numa Pass in Kootenay National Park, in the southern Rocky Mountains. Right: Knight Inlet, an example of the deep fjords that penetrate the Coast Mountains.

Facing page top: McDonald Creek valley in Stone Mountain Provincial Park, in the northern Rocky Mountains. Facing page bottom: Forest and landscape of the Peace River valley.

the arid southern Interior valleys. Grassy alpine tundra and scrubby willow-birch habitats are common in northern BC and at high elevations in southern parts of the province.

BC supports the highest eulipotyphlan diversity (17 species) of any Canadian province, largely due to its environmental diversity. On a provincial scale, the richest area (home to 10 species) is the lower Fraser River basin in southwestern BC. The high eulipotyphlan diversity here can be attributed to the presence of six species found nowhere else in the province (or in Canada): OLYMPIC SHREW, PACIFIC WATER SHREW, TROWBRIDGE'S SHREW, AMERICAN SHREW-MOLE, COAST MOLE and TOWNSEND'S MOLE. All are associated with the coastal lowlands and western slopes of the Cascade Range or southern Coast Mountains. Typically, they range from California through Oregon and Washington to their northern limits in southwestern BC. Their distributions suggest that they evolved in the humid coastal forest region of the Pacific Northwest.

Except for the AMERICAN SHREW-MOLE and the OLYMPIC SHREW, which are found at higher elevations in the Cascades, most eulipotyphlan species in the Fraser River basin are confined to low elevations. The Coast and Cascade mountain ranges and water bodies such as Howe Sound and the Fraser River seem to be major barriers that limit their dispersal. The restriction of the highly fossorial COAST MOLE and TOWNSEND'S MOLE to the bottomlands of the lower Fraser River valley probably reflects the mild winters, rich moist soils and abundant earthworms of this region. The AMERICAN SHREW-MOLE is the only mole found outside this area. It ranges up to 1,380 metres elevation in the Cascade Range of BC, and evidently there are isolated populations in humid valleys on the arid east slopes of the Cascades. Active above ground, it is less restricted by soil conditions.

Outside of southwestern BC, moles are absent and the number of shrew species is similar with no obvious patterns in species diversity associated with latitude or elevation. Although there may be differences in species composition, each ecoprovince of BC typically supports four or five shrew species. Arid grasslands of the Interior valleys and plateaus are not productive habitats for shrews, but two—MERRIAM'S SHREW and PREBLE'S SHREW—appear to be restricted to sagebrush and bunchgrass habitats in southern BC. Both were only recently discovered in BC. Inhabiting the arid grasslands of the Columbia Plateau in Washington, the BC populations define the northern limits of these species' range.

AMERICAN WATER SHREW, CINEREUS SHREW, DUSKY SHREW, VAGRANT SHREW, WESTERN PYGMY SHREW and WESTERN WATER SHREW are boreal or cordilleran species that occupy a broad range of forest and wetland habitats across western Canada. Three peripheral species with limited distributions in the north are TUNDRA SHREW, ARCTIC SHREW and AMERICAN WATER SHREW. TUNDRA SHREW is an Arctic mammal found only as a localized population in extreme northwestern BC widely separated from the nearest populations in

Left to right, top to bottom: Coastal rainforest on Moresby Island in Haida Gwaii. Mature WESTERN REDCEDAR forest in the Skagit valley. Shrub-steppe habitat near Osoyoos Lake in the southern Okanagan valley. Grassland and small lakes near Riske Creek in the Chilcotin.

Alaska and the Yukon. It may be a population that was isolated at the end of the last glaciation. ARCTIC SHREW has a broad range across most of the boreal forest of Canada. In BC, where it is restricted to the Peace River region in the northeast, it is at the western limits of its range. Although known from only two occurrences in BC, AMERICAN WATER SHREW is a boreal species that likely shares a range similar to that of ARCTIC SHREW.

One of the more intriguing aspects of shrew distributions in this province is their occurrence on numerous coastal islands. The colonization of the numerous north Pacific coastal islands by shrews remains one of the mysteries of island biogeography. Poorly insulated with short fur and unable to survive more than a few hours without food, they would be expected to be poor colonizers of islands. With the cold-water temperatures of the north Pacific Ocean and the lack of food on floating debris, shrews would not be expected to survive long either swimming or rafting. Yet, four of our shrews are found on BC islands: CINEREUS SHREW, PACIFIC SHREW, VAGRANT SHREW and WESTERN WATER SHREW. The ubiquitous PACIFIC SHREW is especially widespread, occurring on at least 90 BC islands that range from 0.04 to 33,000 square kilometres in area.

How did tiny shrews reach these islands? Except for parts of Haida Gwaii and a small region on Vancouver Island, coastal BC and its associated islands were covered with the Cordilleran ice sheet. With glacial retreat around 13,000 years ago, mammals recolonized the BC coast. On much of the coast, sea levels initially fell below their current levels in this early postglacial period resulting in land bridges that connected some islands to the mainland or at least reduced the water gaps separating them from the mainland, thus enabling shrews to reach these islands. About 6,000 years ago, sea levels began to rise on most parts of the BC coast, effectively disconnecting islands from the mainland and isolating their shrew populations from mainland shrew populations.

Alternatively, shrews may have colonized these islands more recently, crossing existing water barriers by swimming or rafting on floating logs or debris. In Europe, shrews have been documented to cross water barriers of several kilometres colonizing islands in the Baltic Sea or in freshwater lakes. Their insular shrew populations are dynamic showing a pattern of continual colonization and extinction on islands. A few BC islands supporting shrew populations are highly isolated. However, of the 102 BC islands with known shrew occurrences, 99 (97 per cent) are 3.5 kilometres or less from a source population of shrews. For small islands (less than 0.8 square kilometres) with shrews, the lack of geographic isolation is striking with all less than 1 kilometre from a source population. Genetic studies would help unravel the complex history of BC's insular shrew populations and the extent of genetic exchange among the island and mainland populations.

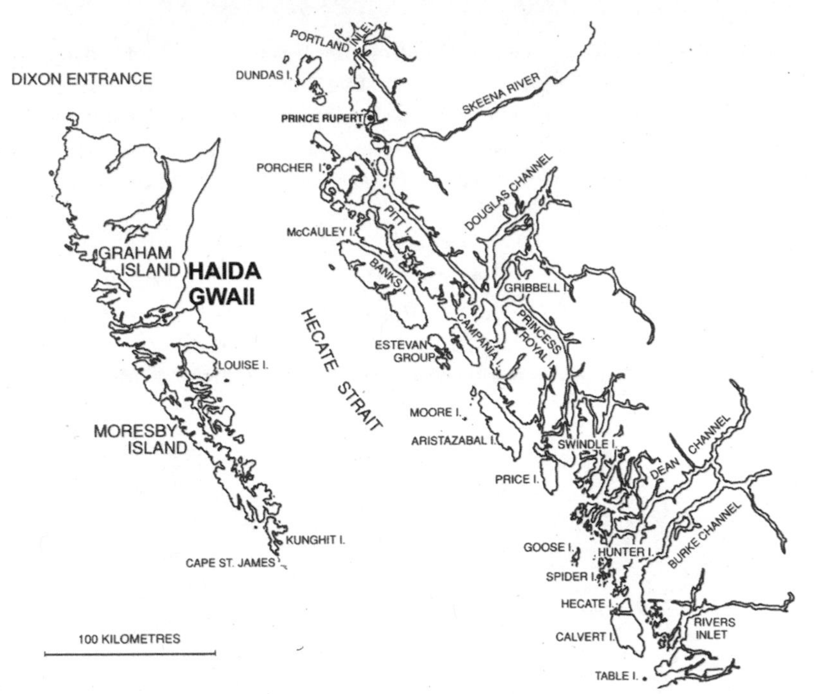

Islands and inlets on the north coast of BC.

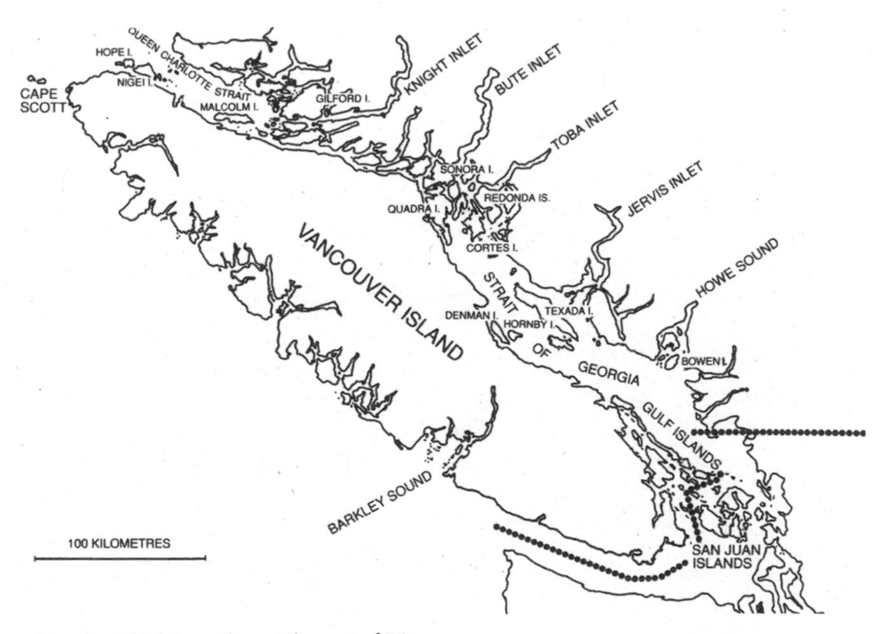

Islands and inlets on the south coast of BC.

Compared with shrews, moles are stronger swimmers and are better insulated with a larger body size. Yet, no moles are native to any BC island, and they occupy only a few of the Washington State islands in Puget Sound. Unsuitable soil or a lack of earthworms possibly accounts for the absence of moles from these islands. (See Relations with Humans, page 36, for a further discussion of earthworm occurrence in BC.)

References

Baker and Bradley (2006); Burgin et al. (2018); Demarchi (2011); Douady and Douzery (2009); Dubey et al. (2007); He et al. (2017); Lloyd and Eberle (2008); McTaggart Cowan (1941); Patton and Conroy (2017); Peltonen et al. (1989); Rzebik-Kowalska (2003); Woodman (2018).

General Biology

Form and Function

As members of the order Eulipotyphla, shrews and moles share a number of similar traits, but they demonstrate major differences in their body structures. Shrews have retained the basic mammalian body plan for locomotion, but their feeding structures, particularly their skulls and teeth, are specialized. Moles, in contrast, developed a highly modified body form for a life underground but retained the primitive dental plan of eutherian mammals.

General Appearance

Shrews are small mammals with long pointed noses. Superficially, they appear rather mouse-like. Their eyes and external ears are small but visible in the fur. Average body mass for BC species range from about 2.5 to 17.5 grams, making them among our smallest mammals. Our only other mammals with similar body mass are bats. Shrews walk on the soles of their feet, and their limbs and feet are not particularly specialized. The exceptions are water shrews. Adapted for swimming, their toes and feet have a fringe of stiff hairs that increase the surface area of the hind feet and assist with propulsion in water.

Shrews are covered with short dense fur. Most tend to be dull brown, but there are several rather attractive species. The dorsal fur of water shrews is nearly jet black; ARCTIC SHREW and TUNDRA SHREW have striking dark bands on the back that contrast with the paler-brown sides creating a saddlebacked pattern. Widespread species such as CINEREUS SHREW demonstrate considerable geographic variation in their fur colour, with populations living in humid coastal regions darker than those living in arid environments of the dry Interior. BC shrews have distinct summer and winter pelages undergoing two moults. Their winter pelage is thicker than the summer pelage and darker in colour. In spring, animals born the previous year that survived over winter shed their winter fur and acquire a summer pelage. Females usually moult before males. Individuals of this age group will reach senescence by 18 months of age and die by early autumn. In autumn, young-of-the-year lose their juvenile summer pelage and acquire winter fur.

Highly modified for digging and moving underground, the mole body is stout and cylindrical, with a short neck and tail. Larger than most shrews, the average body mass for BC moles ranges from about 11 grams for AMERICAN SHREW-MOLE to 140 grams for TOWNSEND'S MOLE. A long snout protrudes well beyond the upper jaw. The eyes are minute and hidden in the fur, and there is no external ear. The front feet are highly modified for digging.

Moles are unique among fossorial mammals because they dig with alternating lateral front-to-back movements of their broad, powerful hands. Other fossorial mammals, such as pocket gophers, typically dig with their front feet held under the body. A mole's front foot is turned permanently outward from the body. It is broad and covered with thick scaly pads of skin; the fingers are equipped with long claws. In highly fossorial moles, such as the COAST MOLE and TOWNSEND'S MOLE, the front limbs cannot be placed directly under the body, and much of the weight is taken by the chest; the palms of the feet cannot be placed flat on the ground, and only the inside edges of the front feet provide support when walking. AMERICAN SHREW-MOLE is somewhat intermediate between fossorial moles and shrews in its locomotion. A capable digger, it also moves on the surface of the ground with considerable agility. It can rotate the front feet under the body and place its palms on the ground, but it actually walks on the backs of the front claws with the front feet bent inward. The hind feet of all moles are adapted for walking, and they resemble those of shrews.

Mole fur is short and velvet-like; it can move in any direction without standing on end, an obvious advantage when moles move backward in their tunnels. Fur colour varies from grey to dark brown or nearly black. Similar to shrews, moles undergo two moults: a spring moult to a summer pelage, and in late summer or autumn a moult to winter pelage. However, winter and summer pelages are less distinct than those of shrews.

Cranial/Dental Traits

The shrew skull is long and narrow with a small braincase. It lacks a zygomatic arch, the arch of bone that extends across the orbit of the eye socket in most mammals. The inner ear bones are not covered by auditory bullae, a bony covering. All BC species have 32 teeth. As is typical of most mammals, shrews have two sets of teeth, but their milk teeth are shed while they are still embryos and thus they are born with their permanent set of teeth. The dentition is different from that of rodents.

In the upper jaw, the front pair of teeth (incisors) are large with a prominent two-hooked cusp. On the middle, front face of this tooth, there is often a tiny accessory cusp called a medial tine. The size and position of the tine is used in shrew identification. The upper incisors are followed by five pairs of smaller

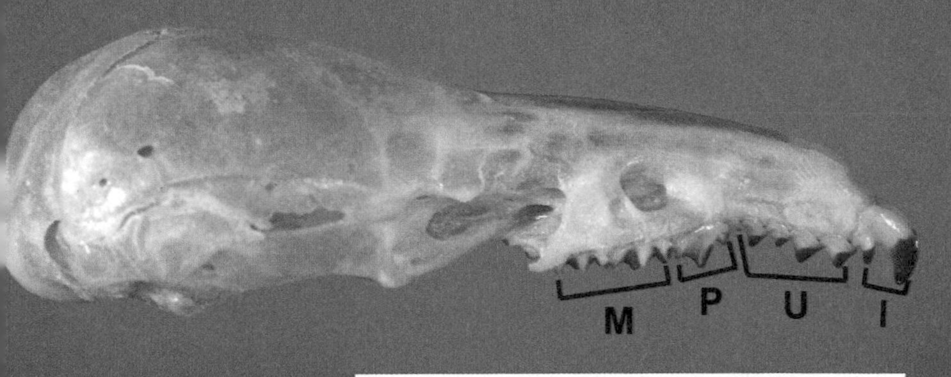

Lateral view of a shrew skull showing the upper incisor (I), five unicuspid teeth (U), premolar (P) and three molars (M).

Ventral view of a shrew skull showing the upper incisor (I), five unicuspid teeth (U), premolar (P) and three molars (M) on each side. Inset: Arrows mark the medial tines on the anterior face of a shrew's upper incisors.

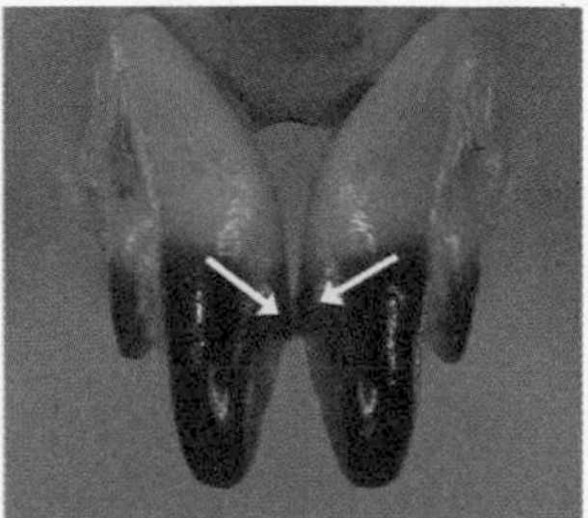

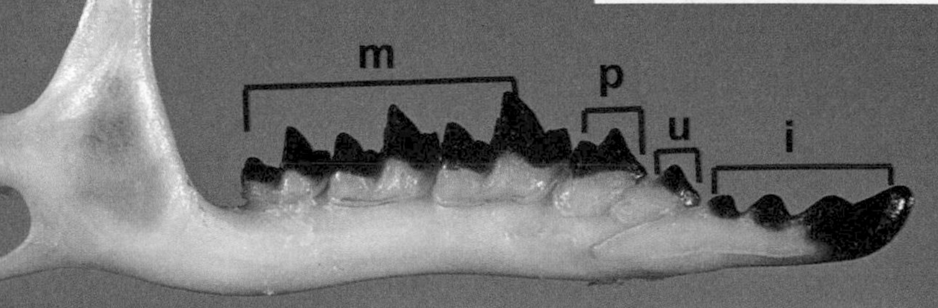

Lateral view of a shrew dentary showing the incisor (i), unicuspid (u), premolar (p) and three molars (m).

teeth that have single-pointed cusps, or unicuspids. Mammalogists cannot agree on which of these teeth are incisors, canines or premolars, and they usually refer to them simply as unicuspids. The upper incisors are always rooted in the premaxillary bone of the skull, whereas the upper canines and premolars are rooted in the maxillary bone. The suture that separates these two bones is not visible in the adult shrew skull, making it difficult to correctly classify the five pairs of unicuspid teeth. The relative size of the unicuspids is an important diagnostic trait for shrew identification. The unicuspids are followed by four pairs of cheek teeth (one pair of premolars and three pairs of molars) that have sharp high-pointed cusps.

The first pair of teeth in the lower jaw (incisors) projects forward; they are followed by a pair of unicuspids. Some mammalogists classify the lower unicuspid pair as incisors, and others consider them to be canines. The unicuspids are followed by one pair of premolars and three pairs of molars. The front incisors act as pinchers for capturing and handling prey; the pointed unicuspids and sharp-cusped cheek teeth are designed for piercing and cracking the hard exoskeletons of invertebrates. The jaw and its associated muscles facilitate rapid biting movements.

Why Red Teeth?

Pigmentation in tooth enamel has been found in a few ancient mammal fossils dating back as far as the late Cretaceous. The earliest red-toothed shrew (Soricinae) fossil dates from the mid-Eocene epoch about 45 million years ago. As members of the Soricinae group, all BC shrews have red pigment on the tips of their teeth. Unlike rodents where the coloration is limited to the incisor teeth, all teeth in the shrew skull and mandible are tipped with pigment. The coloration is created by iron compounds deposited in the outer enamel of the teeth increasing the hardness of the enamel. Possible advantages for having iron in the teeth are to create resistance to acids that may be in food, create sharp cutting edges by differential wear, provide resistance to abrasion and wear, or to prevent cracking. Although shrews eat soft food such as earthworms and insect larvae, some of their invertebrate prey have hard exoskeletons. Even soft prey can be covered with grit or have abrasive material in the digestive tract. More research is needed to test these various hypotheses, but resistance to tooth cracking or abrasion seem to be the most likely explanations. The intensity and distribution of pigment on the face of the upper incisors, the dorsal surface of the upper unicuspid teeth, and the lower incisors differ among species and are somewhat obscure traits used by mammalogists in shrew identification.

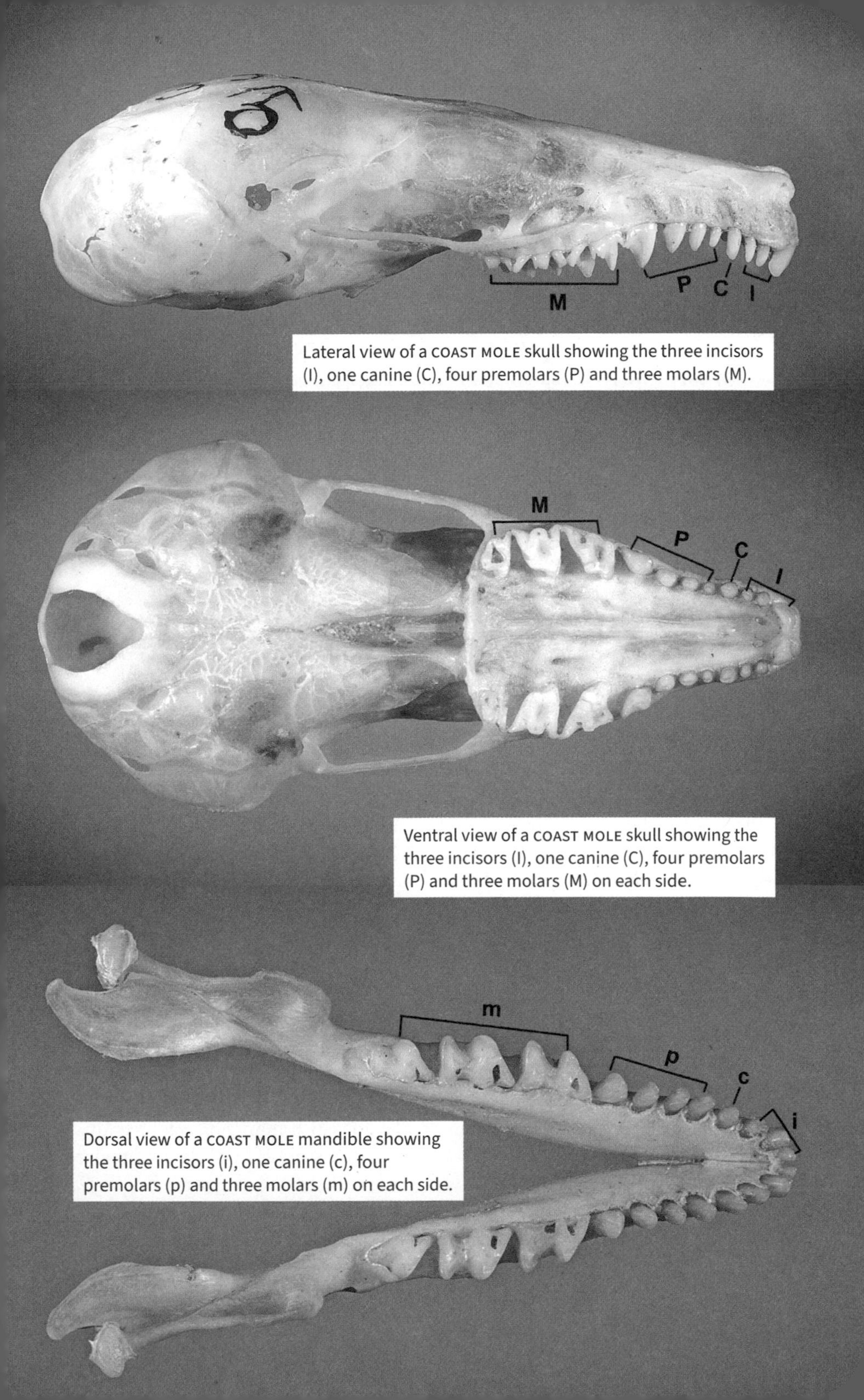

Lateral view of a COAST MOLE skull showing the three incisors (I), one canine (C), four premolars (P) and three molars (M).

Ventral view of a COAST MOLE skull showing the three incisors (I), one canine (C), four premolars (P) and three molars (M) on each side.

Dorsal view of a COAST MOLE mandible showing the three incisors (i), one canine (c), four premolars (p) and three molars (m) on each side.

A mole's skull is somewhat similar to a shrew's with its long narrow rostrum, small braincase and jaw designed for rapid biting. However, the skull has a thin zygomatic arch and the dentition is less specialized. COAST MOLE and TOWNSEND'S MOLE have complete auditory bullae that cover the inner ear bones; the auditory bullae of AMERICAN SHREW-MOLE are incomplete only partly covering the ear bones. The number of teeth in BC species ranges from 36 in AMERICAN SHREW-MOLE to 44 in TOWNSEND'S MOLE and the COAST MOLE. The first pair of upper incisors are flattened on the front and back, and lack the double hook of shrews; the first pair of lower incisors do not project forward. The canine teeth resemble the incisors; the upper molars have distinct W-shaped cusps. Unlike shrews, mole teeth lack red pigment. AMERICAN SHREW-MOLE is born with a set of well-formed milk teeth that resemble its permanent teeth. The milk teeth are functional and are shed when the young are weaned. In contrast, milk teeth of COAST MOLE and TOWNSEND'S MOLE are rudimentary and are resorbed before birth, so they are born with their permanent dentition.

Skeletal Design

Because they are generalized sole walkers, shrews have the basic mammalian skeletal design. Moles, in contrast, demonstrate some remarkable modifications in the forelimb and shoulder girdle that are associated with digging. The elbow joint is positioned so that the upper arm bone (humerus) extends sideways from the shoulder bone (scapula) and the forearm bones (radius and ulna) extend diagonally forward with the palm of the hand facing outward. The power stroke for digging comes from a rotation movement of the upper arm; flexing the elbow only raises and lowers the hand. To support the powerful digging muscles, the shoulder blade and breastbone are elongated and the humerus is short, massive and rectangular.

Sensory Apparatus

Moles and shrews generally live in a world dominated by touch, sound and smell. Shrews have small eyes that can distinguish light and dark, but their ability to detect moving objects is minimal. The anatomy of the retina suggests that shrews may be able to see colours, and colour discrimination was recently demonstrated in a study of the EURASIAN WATER SHREW. In contrast to their limited vision, shrews have well-developed senses of smell, touch and hearing. Sensitive tactile hairs (vibrissae) on the muzzle help them navigate and locate prey in the dark. Shrews rely on smell to locate food and for communication.

They are noted for having a strong musky odour due to numerous scent glands on the body. Most prominent are the flank glands, specialized skin glands on the sides of the body. Flank glands are well developed in breeding males and appear as rectangular areas (five to eight millimetres in length) with short tufts of hairs covered with a sticky secretion. You can see these glands easily by blowing the fur on the side of the body; in old males the skin around these glands is often devoid of hair. Females have indistinct flank glands consisting of reddish patches of capillaries that are only visible on the inside of their skin and can only be seen when the skin is removed.

Despite their small external ears, shrews have well-developed hearing; there is evidence that some species are capable of echolocation. CINEREUS SHREW

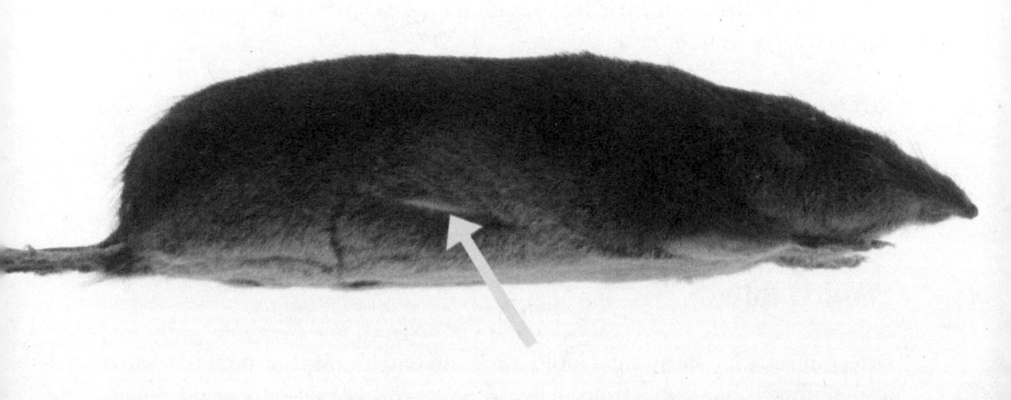

Flank gland on the side of a male WESTERN WATER SHREW.

and VAGRANT SHREW may emit low-intensity high-frequency sounds (18 to 60 kilohertz) for locating obstacles. However, recent research on AMERICAN WATER SHREW found no evidence that it uses sonar underwater or echolocation out of water. Captive COMMON SHREWS from Europe produced twittering sounds (2 to 20 kilohertz) that were used for close-range navigation by echolocation. No ultrasonic clicks similar to those produced by bats were recorded, and rather than detecting prey, the echolocation sounds made by this shrew provide information on nearby obstacles and habitat features.

Eyesight is even less developed in the moles. This is hardly surprising for mammals that spend much of their life in dark underground tunnels. The eyes of COAST MOLE and TOWNSEND'S MOLE can probably distinguish light from dark, but little else. The retina of the AMERICAN SHREW-MOLE's eye is buried under the skin, making it completely blind. In contrast, touch is well developed and the most important sense for orientation in underground tunnels and for locating food. The snouts of most moles including COAST MOLE and TOWNSEND'S MOLE are equipped with thousands of specialized touch-sensitive structures known as Eimer's organs. These minute structures appear as nipple-like bumps on the skin. Each Eimer's organ contains a bundle of sensory nerves that function to provide tactile signals enabling moles to detect small surfaces and textures while foraging or moving underground. Moles also have vibrissae on the snout, head, hands and tail.

The role of hearing and smell in the life of moles is not well understood. The AMERICAN SHREW-MOLE evidently responds to sounds in the frequency range of 8 to 30 kilohertz, but the hearing of the COAST MOLE and TOWNSEND'S MOLE has not been studied. Smell is poorly developed for many moles and probably of little use in locating food. Nevertheless, moles have a number of accessory organs associated with their reproductive organs, and some are thought to produce scents for communication.

References

Branis and Burda (1994); Churchfield (1990); Giacometti and Machida (1965); Gorman and Stone (1990); Hawes (1976); Hutterer (1985, 2005); Lewis (1983); Marasco et al. (2007); Moya-Costa et al. (2018); Pernetta (1977); Reed (1951); Siemers et al. (2009).

Food Habits

Determining what shrews and moles eat is challenging. Most of our knowledge on their food habits comes from analyzing stomach contents in trapped animals. Direct observations of shrews feeding in the wild are rare, and except for distinctive species such as water shrews, reliable identification by observation of most BC shrews is not possible. Above-ground feeding has been observed for some moles. Using a headlamp with red light to minimize disturbance, Vladimir Dinets, for example, was able to observe several instances of COAST MOLE and TOWNSEND'S MOLE feeding at night. However, moles capture the majority of their food underground. Observations of feeding by captive shrews or moles are described in a number of scientific publications, but these observations may not apply to animals living in the wild. We were unable to find a single BC study on the diet of shrews. Glendenning's observations on COAST MOLE's food habits based on captive animals and dissection of trapped animals are the only BC study of mole diets. Consequently, this section is based mostly on data from research done in Oregon and Washington.

Shrews and moles feed primarily on invertebrates. With their different modes of life, they would be expected to exhibit some differences in hunting

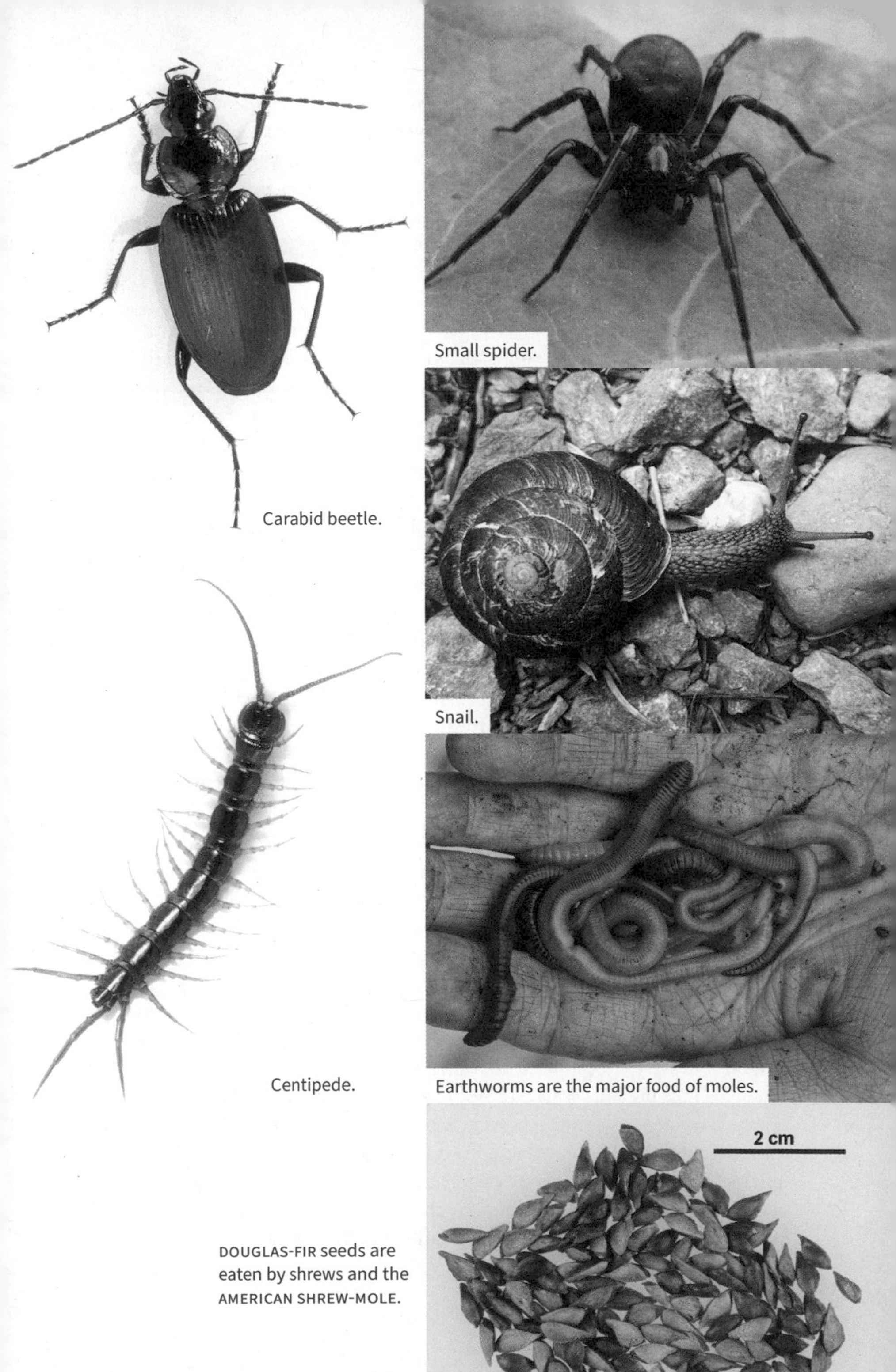

Carabid beetle.

Small spider.

Snail.

Centipede.

Earthworms are the major food of moles.

DOUGLAS-FIR seeds are eaten by shrews and the AMERICAN SHREW-MOLE.

behaviour and diet. Most BC shrews forage in the forest or grassland litter where they feed on subterranean or surface-dwelling invertebrates. There are a few observations of shrews capturing prey above ground in vegetation. The three species of water shrews forage along the edge of streams and ponds capturing aquatic prey underwater. COAST MOLE and TOWNSEND'S MOLE do most of their feeding underground in their tunnels, but AMERICAN SHREW-MOLE is more active on the surface, and much of its hunting takes place in shallow troughs or runways under the litter.

Shrew diets are surprisingly diverse including earthworms, centipedes, spiders, harvestmen, slugs, small snails and many kinds of insects, both adult and larval. Their most common insect prey are moths and butterflies, beetles, ants, grasshoppers, crane flies and true flies. Water shrews specialize on aquatic invertebrates, especially nymphal stoneflies, mayflies and caddisflies. They also prey on small fish and larval salamanders. Captive shrews have been observed eating dead vertebrates such as mice, but the extent to which they scavenge carrion in the wild is unknown. Plant material also turns up in stomach remains. Some is probably ingested accidentally, but fungi, lichens and seeds of coniferous trees are consumed on occasion. Pronounced seasonal variation in diet has been demonstrated for most Eurasian shrews, but winter diets have been described for only a few North American species.

Despite their diverse diets, food habits do vary among species of shrews. Each species seems to have a few preferred prey types that account for a large portion of the diet. Differences in food habits to some extent result from differences in foraging strategies and body size (see Habitat and Community Ecology, page 26). Food habits also vary with habitat and season depending on the abundance of invertebrates. Shrew species also differ in their bite force, with the strongest bite generally found in those species that eat the hardest prey. MERRIAM'S SHREW, a medium-sized shrew that feeds on hard-bodied invertebrates, has a stronger bite than would be expected for its size and is equipped with a particularly robust dentition.

Shrews rely on smell and hearing for hunting. Their echolocation system is not suited for detecting invertebrates, but they may hunt by listening for the sounds made by moving prey. A hunting shrew pounces on and subdues its prey with a series of rapid bites; it may have to pin a larger invertebrate to the ground. Larger prey captured by water shrews, such as minnows, are eaten on land. Once the hunter has immobilized its prey, it eats by passing the food sideways through its mouth, chewing with sharp teeth. Hard body parts, such as legs, wing cases and head, may not be eaten. Small snails are cracked open and eaten whole. Slugs and snails are a problem because their heavy mucus covering often coats the whiskers and fur, which necessitates a good cleaning after a meal. Nevertheless, slugs have been identified in the stomachs of larger shrews.

Are Shrews Venomous?

Although it is common for snakes and spiders to be venomous, few mammals are. Most venomous mammals are members of the Eulipotyphla, with five species confirmed by toxicological studies to produce a weak venom. They include the two species of Caribbean solenodons (Solenodontidae) and three shrews (Soricidae)—two species of Eurasian water shrews and the SOUTHERN SHORT-TAILED SHREW found in the eastern United States. Another 17 shrews including the WESTERN WATER SHREW and CINEREUS SHREW found in BC are suspected to be venomous, based on observations of their behaviour or on their close kinship to known venomous species. Venomous shrews produce a toxic saliva in enlarged salivary glands that is transmitted in a shallow groove located along the inner side of their lower incisor tooth. Experiments on laboratory animals have shown that the weakly toxic saliva affects the central nervous system of their prey, depressing respiration and causing convulsions or paralysis. Of the shrews that have been studied, the SOUTHERN SHORT-TAILED SHREW has the most toxic saliva, and doses administered experimentally to mice were fatal.

The function of venom in shrew biology is contentious, but it presumably helps them subdue large prey. The paralytic effect of shrew venom also enables them to hoard and store comatose prey such as earthworms or insect larvae in food caches for later consumption—an obvious advantage for a small mammal with high energy requirements. Nonetheless, only about 4 per cent of the Eulipotyphla are known or suspected to be venomous. More research is needed to determine the prevalence of this trait among shrews and moles.

BC moles feed mainly on earthworms. Studies done in Oregon found that earthworms represent 40 to 90 per cent of their diet. Other food items eaten include insect larvae, centipedes, snails, slugs and plant material. Prey is located by touch with hairs and Eimer's organs on the mole's snout. A mole usually pins its prey to the ground with the front feet and immobilizes it with a series of bites. It may work an earthworm through its front feet to strip dirt and mucus from the body and squeeze soil from the intestine. Some eat worms from end to end; others simply tear them into smaller pieces before swallowing. Although AMERICAN SHREW-MOLE captures some of its prey above ground, COAST MOLE and TOWNSEND'S MOLE usually forage in shallow surface tunnels. Rather than dig prey from the soil, moles hunt opportunistically, taking invertebrates that live inside tunnels. In southwestern BC, tunnel digging is especially demanding for moles in the dry summer months when the soil is hard.

References

Carraway and Verts (1994); Churchfield (1990, 1994); Churchfield, Rychlik and Taylor (2012); Dinets (2017); Glendenning (1959); Gorman and Stone (1990); Kowalski and Rychlik (2021); Pernetta (1977).

Habitat and Community Ecology

The habitat requirements of shrews are mostly related to invertebrate abundance and physical conditions, such as temperature, moisture, soil type, the amount of leaf litter on the forest floor and the influence of vegetation on environmental conditions. On the continental scale of North America, the diversity of shrews is highest in regions with cool, moist environments and lowest in arid regions. One region of high species richness is the Pacific Northwest. A similar trend is evident for local habitats. Wetlands, riparian areas and moist forests and fields are the habitats that support the greatest diversity and abundance of shrews. Shrews prefer these environments because their high metabolic rates create high moisture requirements and they can easily become dehydrated. Moist environments also tend to have diverse and abundant invertebrate faunas offering a rich food supply.

Most small-mammal communities in BC support multiple species of shrews. Shrews often represent 30 to 40 per cent of the total small-mammal species in a community. The most diverse shrew communities are in the lower Fraser River valley where as many as five species may co-occur: OLYMPIC SHREW, PACIFIC SHREW, PACIFIC WATER SHREW, TROWBRIDGE'S SHREW and VAGRANT SHREW. Dry DOUGLAS-FIR forests of the southern Interior may support four species: CINEREUS SHREW, DUSKY SHREW, WESTERN PYGMY SHREW and VAGRANT SHREW. Information on grassland shrew communities in BC is limited, but three species were taken at the same trap station in shrub-steppe habitat on Mount Kobau: DUSKY SHREW, PREBLE'S SHREW and VAGRANT SHREW. High species diversity is limited to ideal habitats; for most habitats in the province, two or three species is typical. Except for Vancouver Island and some islands in the Great Bear Rainforest, shrew communities on most BC islands consist of a single species: either the PACIFIC SHREW or VAGRANT SHREW.

How do different shrews coexist in a community? A number of factors help reduce competition among coexisting species and permit them to share habitats. In most communities, the majority of the shrew population is represented by one or two common species that often account for as much as 70 per cent of the shrews captured. Ecologists often classify animals as habitat specialists or generalists. Most BC shrews have rather broad habitat requirements and could be called generalists. It is the habitat specialists such as the water shrews that tend to be the rare species in communities.

It appears that in most habitats, the smallest and largest shrews are rare and the intermediate-sized species are the most abundant. Although shrews have flexible diets, their size and bite force determines, to some extent, the prey they take. Species with the strongest bite are better able to exploit hard-bodied prey, such as snails and insects with hard exoskeletons. There may be considerable overlap in types of prey taken by various species, but the proportions of prey types are different. Different foraging strategies also reduce competition. The aquatic specializations of water shrews give them access to prey unavailable to other shrews. Because of its activity in tunnels under the humus layer, TROWBRIDGE'S SHREW exploits more soil-dwelling invertebrates than other shrews.

In addition to differences in abundance, size and diet, differences in microhabitat enable similar species to coexist in a community. Two similar species found together over much of southwestern BC are PACIFIC SHREW and VAGRANT SHREW. In an elegant study, Myrnal Hawes demonstrated subtle differences in their use of habitat. PACIFIC SHREW appears to be a forest species, whereas VAGRANT SHREW is associated with grassy fields and rich moist soils near water. In a one-hectare study plot near Maple Ridge, PACIFIC SHREW was most common in WESTERN HEMLOCK forest with acidic soils and VAGRANT SHREW was associated mainly with the less acidic soils of WESTERN REDCEDAR forest. The two species have a similar bite force. However, the PACIFIC SHREW has more robust and durable teeth that are less prone to wear, which may make it better adapted to feed on the small, hard-bodied invertebrates associated with acidic soils. VAGRANT SHREW, on the other hand, is better adapted to the less acidic rich soils where there are more soft-bodied prey, particularly insect larvae.

The habitat requirements of moles are closely linked to soil types. Soil conditions (texture, moisture) and earthworm abundance are particularly important for the highly fossorial COAST MOLE and TOWNSEND'S MOLE. Although they can be found in the same habitat, TOWNSEND'S MOLE prefers drier well-drained soils than the COAST MOLE. Known occurrences of TOWNSEND'S MOLE in BC are restricted to Marble Hill and Ryder soil types that occur in a small area near Abbotsford in the Fraser River valley. AMERICAN SHREW-MOLE is not so closely tied to soil conditions, and this is reflected in its broader geographic range. Although it shares the same habitats with a number of shrew species, competition is likely reduced by its greater subterranean activity.

References

Churchfield (1994); Hawes (1975, 1977); Huggard and Klenner (1998); Nagorsen et al. (2001); Sheehan and Galindo-Leal (1996); Wrigley, Dubois and Copland (1979); Zuleta and Galindo-Leal (1994).

Reproduction, Development of the Young and Longevity

Shrews rarely live beyond 18 months. Females of some species may breed in their first summer. Typically, however, most breed in the spring following the year of their birth and then die. Their breeding strategy is to produce as many young as possible in a single breeding season. Females produce large litters and have the potential to produce several litters in the breeding season. Moles, on the other hand, may live three or four years, and their breeding effort is extended over several successive breeding seasons. Female moles have only a single litter each year, and they produce fewer young.

All shrew species in BC have a distinct breeding season that begins in late winter or early spring. The general cue for reproductive activity is the change in the daily light cycle. But the onset of breeding is extremely variable among species and geographically among different populations of a species. In BC, WESTERN WATER SHREW and TROWBRIDGE'S SHREW seem to be the earliest breeders with pregnant females found as early as February. Males reach sexual maturity several weeks before females. Geographic variation in the onset of breeding can be substantial; the breeding season is delayed in populations living at high latitudes or high elevations. This variation ensures that the young are born at a time when there are maximum food resources and the climate is mild.

Little is known about courtship and mating of shrews. Estrus, the time of the reproductive cycle when females are receptive to males, lasts only a few hours. Scent probably plays a major role in ensuring that the sexes meet at the appropriate time. Odours from the flank glands of males may have some role in communication among the sexes. The gestation period is 24 to 25 days. Because of the difficulties in observing wild shrews, information on litter size is based on counts of embryos found in the uterus. As many as 12 embryos have been reported for BC species, but a typical litter is 5 or 6. Females may come into estrus immediately after giving birth, and it is possible for a shrew to become pregnant while nursing a litter. Most species are capable of having two litters in the breeding season, and some may produce three. The number of litters varies among different populations and even from year to year within a population. To some extent it is determined by the survival of adult females. Only females that manage to survive predation or starvation and until late summer can produce three litters.

Pregnant females construct nests of dried grass and leaves in underground burrows or in sheltered locations under logs or vegetation. The brief period of parental care falls solely on the mother; males take no part in the raising of the young. Shrews are born in a more immature state than the young of other eutherian mammals. Newborns weigh less than one gram and are underdeveloped. Their skin is pink, the eyelids are closed, the ears are fused to the back of the head and the digits are clawless. By the sixth day, the skin becomes pigmented, and at around eight days the hair begins to erupt. The eyes and ears open in 14 to 18 days. The permanent set of teeth are present at birth but do not erupt through the gums until about two weeks later. Young shrews grow and gain body mass rapidly, and by three weeks of age they are full adult size. By this time, they are weaned and soon become independent. Generally, young shrews do not breed until the following spring, but first-year breeding does occur in the females of some species. Myrnal Hawes, in her study of VAGRANT SHREW in southwestern BC, observed that as many as 50 per cent of the females in some populations bred in their first summer. It appears to be females born early in the year that manage to breed in the summer of their birth.

Shrew Caravans

Young shrews remain with their mother until they are weaned at about three weeks of age. During this phase of their life, if their nest is disturbed or the family is exploring their surroundings, the young shrews may move about by caravanning. They form a line behind their mother, with each touching the rump or holding the base of the tail of the preceding shrew with the mother leading the conga line. Most observations of caravanning are for species of Eurasian shrews, but it has been observed in captive young CINEREUS SHREWS and PACIFIC WATER SHREWS in the wild.

Glenn Ryder, an astute BC naturalist, had the good fortune to observe and record his observations of caravanning in a family of PACIFIC WATER SHREWS in the lower Fraser River valley:

> Bendire or Pacific water shrew family out at the east side of the beaver pond on Davidson Creek. Female in lead with her 6 young trailing behind her with the lead young holding on to the tip of the female's tail and each young had a hold of the young in front of it. All young had a tail tip in mouth until they reached a big log. They let go and all young stayed put while the female was diving down into the pond water at cattail clumps getting food for her young. She made 7 trips feeding all the young and her last trip was food for herself (likely dragon fly nymph, etc.). After the meal they all went up the hillside to the east woods holding tail tip to tail tip and vanished in the vegetation and forest floor litter.

(From "Field Observation of Caravanning by a Family of Pacific Water Shrews in British Columbia," *Wildlife Afield* 7, no. 2 (2010): 298–300.)

Four shrews less than a week old in a grass nest, Delta, BC.

Survivorship in young shrews is low; less than half will survive beyond five months, and only a few will survive long enough to breed. After leaving the nest, shrews have to find a suitable area to establish and maintain a home range. Old shrews with established territories exclude young shrews from the most productive habitats, and many dispersing young probably starve before they can establish a territory. Predators also take their toll. Owls and hawks prey readily on shrews, whose remains are common in their pellets. However, because of their pungent odour, shrews may be unpalatable to many mammalian carnivores and they are not a major food item. Dead shrews with bite marks from predators such as ERMINE or DOMESTIC CATS are often found discarded on trails or roadsides. In BC, shrew remains have been recovered from the stomachs and scats of ERMINE, PACIFIC MARTEN and AMERICAN MARTEN.

Wild adult shrews rarely survive beyond their second summer. Few can cope with the energy drain of the breeding season followed by the stress of food shortages in autumn. There are a few exceptions though. In one of her study areas, Myrnal Hawes observed a VAGRANT SHREW that survived two winters. Shrews in captivity may live beyond 18 months.

In late spring and summer, shrew populations consist of two distinct generations: old adults born the previous year and young-of-the-year. Several traits distinguish these groups. An old adult has badly worn teeth with little

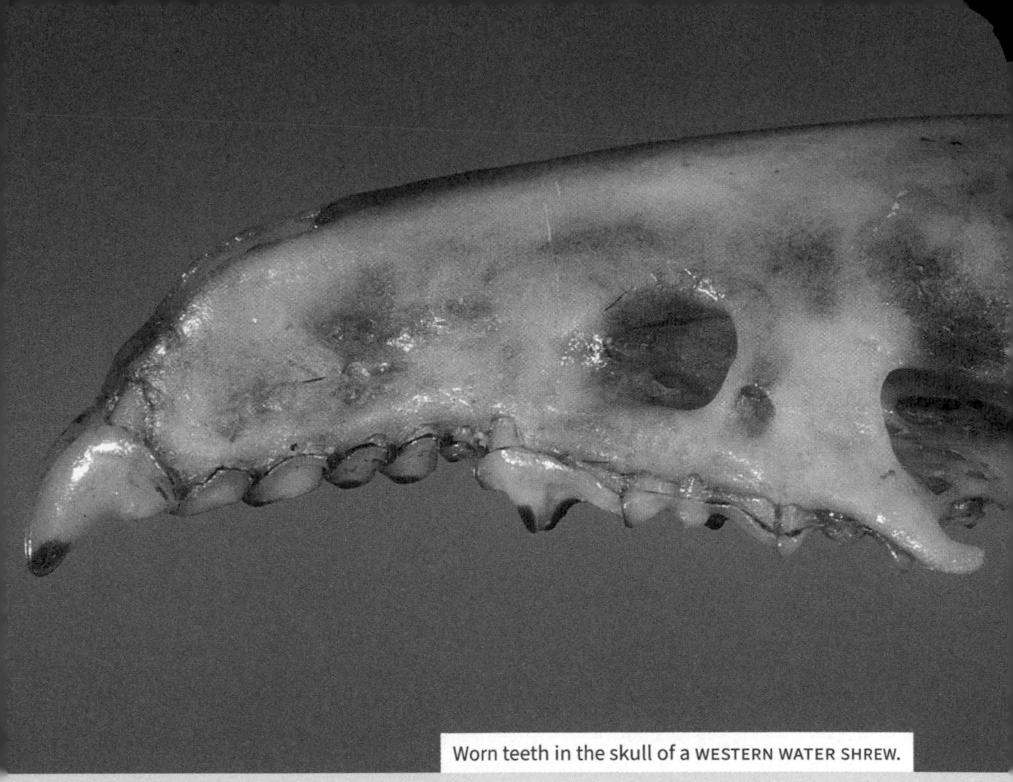
Worn teeth in the skull of a WESTERN WATER SHREW.

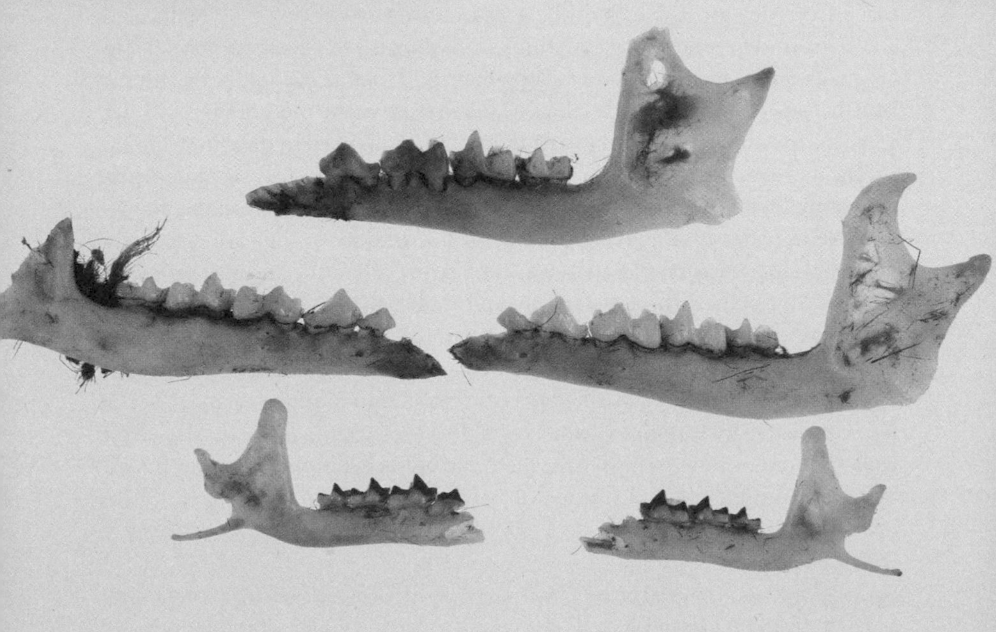
Dentaries of AMERICAN SHREW-MOLE and VAGRANT SHREW found in BARN OWL pellets.

pigmentation, a flattened braincase and nearly hairless tail and feet. A young shrew has unworn teeth with heavy pigmentation, a domed braincase, and well-furred tail and feet with a distinct tuft of hairs on the tip of the tail.

Despite living in the darkness underground, moles also have a well-defined breeding season. Presumably, their occasional forays to the surface provide them with adequate cues associated with changes in the daily light cycle. The breeding season of COAST MOLE and TOWNSEND'S MOLE begins in early January when the males' testes begin to enlarge. The first young are born in April. AMERICAN SHREW-MOLE has a more prolonged breeding season, although most births occur between March and May. Little is known about the reproductive behaviour of moles in North America. Male EUROPEAN MOLES become very active during the breeding season, travelling long distances and constructing many new tunnels in an attempt to enter the tunnel system of a receptive female. Mating evidently takes place underground. Precisely how the sexes locate each other underground is not clear, but communication by scent produced by various skin glands or the perineal glands of males seems most likely.

The gestation period has not been determined for the three species of BC moles, but it is likely four to six weeks. The litter size ranges from one to four, with three young most common. COAST MOLE and TOWNSEND'S MOLE construct nursery nests in underground chambers for rearing their young. The nests of TOWNSEND'S MOLE are particularly elaborate, with inner and outer layers made from dry grass. Nests of the AMERICAN SHREW-MOLE may be located above ground in hollow logs or stumps. Newborns are naked and blind, and lack nails on the feet. They also lack pigment and their skin is pink. By about a month, the young are fully developed, and they leave the mother's nest. Moles reach sexual maturity at about 10 months of age and breed in their first winter.

The highest mortality in moles occurs in late spring and early summer when the young leave the nursery nest. Many probably never find a suitable home. Because they disperse above ground, this is a time when they are extremely vulnerable to predators. TOWNSEND'S MOLE and COAST MOLE are most prevalent in BARN OWL pellets from May to July, and almost all of the skulls recovered in pellets are of immature moles. The maximum life span of moles ranges from four to six years, depending on the species. Based on progressive wear of the teeth, Valentin Schaefer classified COAST MOLES into four age categories. His oldest age category was three to four years. However, using a technique based on counting annual growth rings in the teeth, biologists in Europe have determined the maximum life span of EUROPEAN MOLES to be six years.

References

Forsyth (1976); Gorman and Stone (1990); Gusztak and Campbell (2004); Hawes (1975); Kuhn, Wick and Pedersen (1966); Ryder (2010); Ryder and Campbell (2007); Schaefer (1978); Sealy (2017).

Conserving Energy and Winter Survival

There is a cost to being small. Small mammals tend to have relatively high metabolic rates for their body size resulting in large energy requirements. Because shrews and moles are lean mammals with limited fat reserves, they live in a precarious balance where their energy stores are continuously threatened by exhaustion. To avoid starvation they must feed often.

This is especially acute for shrews because most cannot survive more than three hours without food. Some small mammals, such as bats, have the ability to enter torpor at times when food is scarce—that is, they simply lower their metabolic rate and reduce their body temperature for brief periods, thus reducing their energy expenditures. None of our shrews can enter torpor; they have to maintain a constant body temperature of around 38.5°C.

Nevertheless, they have a number of strategies to avoid starvation. Voracious eaters, they typically consume their own body mass in food each day. Most are opportunistic feeders with flexible diets—thus, they can exploit whatever prey happens to be available. Captive shrews hoard excess food for later consumption, but it is not known if this behaviour is prevalent in wild shrews. Food caching would be an important adaptation to ensure a constant food supply during times of high energy requirements.

To conserve energy, shrews rely on the insulation provided by their fur and nests to reduce heat loss and lower their energy demands. In northern temperate regions, winter is one of the most demanding times in the life of a shrew. Unlike bats, shrews do not hibernate but remain active throughout winter. Well-insulated nests and insulation from snow provide some protection from extreme temperatures. Most shrew activity occurs under the snow in the subnivean space (between the snow and the ground surface). When snow depths reach 15 to 20 centimetres, this space provides an environment with nearly constant temperature and humidity. Temperature differences between the subnivean area and the exposed surface of the snow may be as great as 25°C. Periods when the snow cover is sparse, such as early winter and spring, are the most critical. In coastal regions of southwestern BC, winters are mild and shrews are rarely exposed to the temperature extremes of northern Interior regions. Nevertheless, the occasional flooding or a brief cold spell could be stressful for them.

Shrews curtail their activity in winter, and feeding periods tend to be brief, with much time spent in their well-insulated nests. Evidently, a number of winter-active invertebrates are available under the snow, ensuring a winter food supply. Although most are small (one to four millimetres), they tend to be energy rich. Many shrews consume seeds in winter, although the importance of seeds in the winter diet is not clear. Seeds may be a high-energy food or may be used as an alternate food only when invertebrates are scarce. Further evidence for the importance of food is the fact that many shrews establish winter territories to ensure they will have access to sufficient food resources. Myrnal Hawes found that PACIFIC SHREW and VAGRANT SHREW in southwestern BC established winter territories in autumn that they retained until the onset of breeding the following spring.

Besides behavioural adaptations, shrews undergo some major physiological changes that enable them to cope with winter. Young-of-the-year undergo a fall moult in which the short juvenile fur is replaced with a long, plush winter pelage. Experimental studies have shown that the winter fur of shrews is about 30 per cent more effective at retaining heat than their summer fur. Some accumulate brown fat, an important heat source for mammals. An unusual feature of shrews is heavy deposits of brown fat in the limbs, which probably warms the blood returning from the poorly insulated extremities.

Shrinking Shrews

Perhaps the most unusual winter adaptation of shrews is a shrinkage in overall size. This adaptation is often termed the *Dehnel phenomenon*, in honour of the Polish biologist August Dehnel who first described it in the 1940s. In winter a shrew becomes lighter and shorter. The weight loss is not simply a loss of fat, because organs such as the kidneys, liver and spleen become lighter and the volume of the brain actually shrinks. The skeleton is also remodelled: the skull height decreases and the discs between the vertebrae shrink. The Dehnel phenomenon seems to be most pronounced in regions with severe winters. Most research on it has been done in Europe, and it is not known how widespread this adaptation is in North American shrews. Myrnal Hawes, who tracked individual body mass changes in PACIFIC SHREW and VAGRANT SHREW living in the mild coastal climates of southwestern BC, found that an individual's body mass was lowest in the winter months (November to February), but their body length and skull height decreased only slightly in winter. The Dehnel effect would be expected to be more pronounced in shrews living at higher latitudes in northern BC. The general consensus among biologists is that a smaller size lowers energy demands (food requirements) at a time when available food is restricted to small arthropods, and larger prey such as earthworms are scarce. A recent study of European Moles in the Czech Republic by Lucie Nováková and colleagues revealed that moles' skulls may shrink as much as 11 per cent in winter, suggesting that some moles exhibit the Dehnel phenomenon.

Two other stressful periods in the life of a shrew in terms of energy are reproduction and moulting. The breeding season coincides with the time when food resources are most plentiful and temperatures are mild. The spring moult, in which the winter pelage is shed and replaced by the summer fur, is closely linked to the breeding season. In BC, female VAGRANT SHREWS and PACIFIC SHREWS complete their moult in late winter, a week or two before the onset of breeding, thus avoiding any overlap in these energy-demanding activities. Oddly, the sequence is reversed in males, with the onset of reproductive activity taking place in winter and the moult delayed until May or June after the peak of sexual activity.

Another consequence of a high metabolic rate is a high loss of water through respiration. Even in humid environments, shrews have high water requirements. They are also unable to withstand high temperatures. CINEREUS SHREW, for example, is severely stressed when the ambient temperature exceeds 24°C.

Moles do not hibernate, and they lack large fat stores. There is no evidence that TOWNSEND'S MOLE and COAST MOLE can enter torpor. Much larger than shrews, their relative metabolic rates are lower, resulting in less extreme energy requirements as those of shrews. They also have denser, longer fur than shrews and should be better insulated. Because they spend much of their life in sheltered tunnels underground, fossorial moles are buffered from the daily and seasonal temperature variations that occur above ground. In the Fraser River basin of southwestern BC, where winters are mild, they presumably burrow below the frost line during the occasional cold spell. Flooding is the most serious challenge for these animals in winter.

In autumn and winter, EUROPEAN MOLES store large quantities of food (usually earthworms) in their tunnels. Some of these caches are enormous, containing as much as 1.6 kilograms of earthworms. Food caching, however, has not been reported for any of the moles in western North America.

In contrast, AMERICAN SHREW-MOLE is smaller in body mass than COAST MOLE or TOWNSEND'S MOLE, with a much higher resting metabolic rate and higher body temperature that closely matches that of a shrew of similar body mass. They have a voracious appetite, and captive AMERICAN SHREW-MOLES consumed more than their body mass in a 12-hour period. When exposed to cool temperatures, its body temperature may gradually cool, and two individuals in a study by Kevin Campbell and Peter Hochachka were observed to undergo a slight reduction in their metabolic rate. More research is needed to establish if this mole uses some form of mild torpor.

Living Underground

Not unlike the environment encountered by humans working underground in mines, the atmosphere in mole tunnels presents physiological challenges. The most energy-demanding activity in the life of a mole is excavating tunnels underground. The soil has to be sheared from the tunnel face with their front feet and then pushed along the tunnel to the surface above ground. The COAST MOLE does most of its digging and tunnel construction in autumn and winter when the soil is moist and soft. Digging not only requires a lot of energy, but it is done in an atmosphere with low oxygen and high carbon-dioxide levels. Valentin Schaefer demonstrated that COAST MOLE tunnels have carbon-dioxide levels 10 times (sometimes 200 times) as high as atmospheric levels above ground. Oxygen levels in the tunnels of this species are also significantly lower than the surface air. Compared with non-burrowing mammals, the oxygen-carrying capacity of TOWNSEND'S MOLE and COAST MOLE blood is higher. Myoglobin, the protein in muscle that stores oxygen, is most concentrated in the muscles of their forelimbs, the limbs that do the heavy work digging.

References

Aitchison (1987a, b); Campbell and Hochachka (2000); Churchfield, Rychlik and Taylor (2012); Genoud (1988); Gorman and Stone (1990); Hanski (1994); Hawes (1975); Hyvärinen (1994); Ivanter (1994); McIntyre, Campbell and MacArthur (2002); Nováková et al. (2022); Schaefer and Sadleir (1979).

Relations with Humans

Ancient Egyptians revered shrews, and their embalmed and mummified remains have been found in animal cemeteries in the Nile valley. In contrast, the ancient Romans considered them evil and dangerous, a view shared in Western folklore. The word *shrew* was used to describe a bad-tempered or aggressive woman—a theme captured in William Shakespeare's play *The Taming of the Shrew*. Today, most humans likely think of them as high-strung, ferocious little creatures, but otherwise regard them with indifference and little appreciation of their natural history.

A few, such as TROWBRIDGE'S SHREW, are considered potential pests in the forest industry because they consume seeds of commercially important trees such as DOUGLAS-FIR. However, any negative impact from eating tree seeds is balanced by their role as predators of insect pests. CINEREUS SHREW and ARCTIC SHREW are major predators of forest pests such as the LARCH SAWFLY. Sawflies are members of the Hymenoptera, an insect order that includes ants, bees and wasps. Adult sawflies, which resemble tiny wasps, lay their eggs on the stems and shoots of their host tree; LARCH SAWFLY is specific to TAMARACK and WESTERN LARCH tree species found in the BC Interior. Their eggs hatch into small caterpillar-like larvae that feed on the needles of the tree. After feeding, the larvae fall to the forest floor, where they spin a cocoon and overwinter until the following spring when they emerge as adult sawflies. When there are large outbreaks, LARCH SAWFLY causes extensive defoliation that reduces tree growth; if attacked repeatedly, trees may eventually die.

Shrews feed on sawfly pupae on the forest floor. Their small size, acute sense of smell and ability to forage in the forest litter make them skilled hunters of the pupae. A single shrew may consume a hundred in a day, and it may destroy even more by damaging hoarded cocoons that are not eaten. During sawfly outbreaks, shrews may be an important control on their populations. In the 1950s, CINEREUS SHREW was deliberately introduced to Newfoundland, where there are no native shrews, as a biological control agent for the LARCH SAWFLY. Larvae of SPRUCE BUDWORM have been recovered in the stomachs of CINEREUS SHREWS, and it is likely that shrews eat other insect pests.

Because of their tunnelling activities, moles tend to have a more negative image than shrews and are generally viewed as pests. Of the three mole species found in BC, only COAST MOLE could be regarded as a pest. It is listed on schedule B of the BC Wildlife Act, which means it can be captured or killed to protect property. Widespread in the agricultural land of southwestern BC, it can occur in large numbers. AMERICAN SHREW-MOLE is mainly a forest species, and TOWNSEND'S MOLE, although an agricultural pest in Oregon and Washington, is too localized in BC to be of any economic consequence on a provincial scale.

The economic damage caused by COAST MOLES in BC has not been measured. In agricultural areas of the lower Fraser River valley, molehills may cause some loss of grazing land; bulb growers, vegetable gardeners and berry farmers often complain of mole damage. Moles may eat bulbs and roots, or damage plants indirectly by their tunnelling activities. Although the effect is mainly visual, mole diggings in golf courses and lawns are also considered a nuisance.

Many control methods have been explored in the long battle with moles. A cursory scan of the internet will turn up a multitude of recommendations on how to control moles with traps, poisons, chemical repellents, noise, flooding and fertilizers. Several types of kill traps are available. We recommend that you consult with provincial regulating agencies for advice on appropriate traps. Several live traps are designed to capture moles alive, but they are not practical for large-scale control in agricultural areas. Poisoning is not recommended; poison baits may be eaten by other animals including pets, and some poisons, such as strychnine, are inhumane. Dead moles contaminated from poisons could be picked up by owls, raptors and various mammals. Castor oil, naphthalene flakes and various commercial products are promoted as repellents, but there are mixed reviews of their effectiveness. Landowners within the range extent of TOWNSEND'S MOLE, an endangered species in Canada listed under the Species at Risk Act (SARA), should check with provincial regulating agencies before undertaking any mole control.

In the past, the short velvety fur of EUROPEAN MOLES was in great demand for clothing, and it was intensively trapped for its pelt. In the 1920s some 12 million moleskins were being imported annually into the United States from Europe. Attempts by the United States Department of Agriculture to create a market for North American moles never succeeded. Today mole fur has fallen out of fashion, and there is little demand for their skins.

An ironic twist to the humans-versus-moles story is the impact of introduced earthworms. Despite the prominence of earthworms in the diet of moles, they are a recent food source brought from Europe in historical time. Only a few earthworm species are native to BC. They inhabit Vancouver Island, and one species occurs also on Haida Gwaii (no moles occur on any of these islands). Evidently, none of the earthworms living in the Fraser River valley, where moles are common, are native species. During his study of COAST MOLES, Valentin Schaefer collected 12 species of earthworms in soil samples from various fields in the Fraser River valley; all were species introduced from Europe.

An intriguing question is, what was the distribution and abundance of moles in the province before Europeans brought earthworms and cleared the forests for farming? With its diverse diet, ability to hunt above ground, and association with non-cultivated land, AMERICAN SHREW-MOLE was able to exploit other invertebrates. However, COAST MOLE and TOWNSEND'S MOLE are more dependent on earthworms, and the high densities of the widespread COAST MOLE seem to be closely linked to earthworm abundance and cultivated farmland. Although the COAST MOLE could have eaten insect larvae and grubs instead of earthworms, it is unknown how large a population this prey base could have supported compared with present-day mole populations.

References

Buckner (1966); McKey-Fender, Fender and Marshall (1994); Schaefer (1978).

Conservation and Threats

Globally, the International Union for the Conservation of Nature (IUCN) lists 50 species (11 per cent) of Eulipotyphla as Critically Endangered or Endangered. Another 26 species (6 per cent) are listed as Vulnerable. The population trend for most of these at-risk species is unknown, but nearly a quarter are thought to be declining. Most of these shrews and moles listed by the IUCN inhabit Africa or southeast Asia, hot spots for Eulipotyphla species diversity, with many endemic to islands. Their threats vary, but a common theme for most is loss and fragmentation of their habitat.

Conservation Status of BC Shrews and Moles

Species	COSEWIC listing	BC listing
Moles		
American Shrew-mole	Not assessed	Yellow
Coast Mole	Not assessed	Yellow
Townsend's Mole	Endangered[a]	Red
Shrews		
Arctic Shrew	Not assessed	Yellow
Pacific Water Shrew	Endangered[a]	Red
Cinereus Shrew	Not assessed	Yellow
Western Pygmy Shrew	Not assessed	Yellow
Merriam's Shrew	Not assessed	Red
Western Water Shrew	Not assessed	Blue[b]
Dusky Shrew	Not assessed	Yellow
Pacific Shrew[c]	Not assessed	
American Water Shrew	Not assessed	Blue
Preble's Shrew	Not assessed	Red
Olympic Shrew	Not assessed	Red
Trowbridge's Shrew	Not assessed	Blue
Tundra Shrew	Not assessed	Red
Vagrant Shrew	Not assessed	Yellow

a Also listed as endangered in the Species at Risk Act (schedule 1).
b Applies to the Vancouver Island subspecies.
c The BC Conservation Data Centre does not recognize PACIFIC SHREW as a distinct species from the DUSKY SHREW.

In Canada, the Committee on the Status of Wildlife in Canada (COSEWIC) has assessed three species of Eulipotyphla. TOWNSEND'S MOLE and PACIFIC WATER SHREW, species found in BC, are ranked as Endangered, and the EASTERN MOLE as Special Concern. Provincially, as assessed by the BC Conservation Data Centre, five shrews and a mole appear on the BC Red List, and three shrews are on the Blue List. Seven of the Red/Blue-listed shrews have yet to be assessed by COSEWIC because of inadequate data or because they are secure in other parts of Canada. They are considered of conservation concern in BC because of their limited distributions, possible rarity, and habitat threats.

In contrast to other small mammals such as bats where there is extensive ongoing inventory and research in the province, shrews and moles are largely ignored. Consequently, the greatest obstacle to determining their conservation status is a lack of information. Because of their listing on schedule 1 of the federal Species at Risk Act, recovery strategies were developed for the endangered PACIFIC WATER SHREW and TOWNSEND'S MOLE. Nevertheless, the strategies are limited by inadequate surveys to determine their provincial range and population trends. Most of the surveys for PACIFIC WATER SHREW are short-term environmental assessments done for development projects in localized areas. The TOWNSEND'S MOLE recovery strategy, including the mapping of critical habitat, is based on a survey of active territories done 19 years ago, and this species's current population and the current condition of its critical habitat are unknown. For the other shrews on the province's Red and Blue Lists, more information on their current geographical range, habitat use and population trends is required to determine their conservation status. Their range maps are largely derived from historical museum specimens. Two listed species are known only from a few historical museum specimens collected more than 40 years ago.

Threats to Shrew and Mole Species

BC shrew and mole populations are impacted by a number of generic threats such as exposure to pesticides or toxic chemicals, predation by DOMESTIC CATS, and poisoning or trapping. However, alteration and loss of habitat are likely the greatest threats.

In western North America, much interest has been focused on the harvesting of old-growth forests and its impact on vertebrate communities; and a number of studies have examined the distribution of shrews and the AMERICAN SHREW-MOLE in forest stands of various ages. Habitat specialists, such as PACIFIC WATER SHREW, are vulnerable to forestry and agricultural operations that affect water quality and aquatic invertebrate communities. However, most of our shrews have rather broad and flexible habitat requirements and should be able to occupy forests of various ages. In the coastal forests of southwestern BC, a study by Druscilla and Tom Sullivan found that shrews seem to prefer forested habitats over recently logged areas with shrew numbers consistently higher in forested habitats. But the response to logging is dynamic. Immediately after clear cutting, shrew numbers in the clearcuts may increase, but several years later, populations stabilize at levels lower than those found in uncut forest. Shrew populations will initially decline in habitats that are logged and burned, but they recover quickly, and by the second year they reach the numbers found in unburned clearcuts.

It is the permanent degradation and loss of habitat from development that poses the greatest threat to BC shrews and moles. Four species of conservation concern (PACIFIC WATER SHREW, TROWBRIDGE'S SHREW, OLYMPIC SHREW and TOWNSEND'S MOLE) are associated with low elevations in the lower Fraser River basin (the Fraser Lowland and Northwestern Cascade Ranges ecosections). Since European settlement in the mid-1800s, there has been a continuing loss of mature forest ecosystems and wetlands to clearing for agricultural land and urban development. South of the Fraser River, forested and wetland habitats have become small patches isolated by development and highways. In the small area occupied by TOWNSEND'S MOLE, agricultural lands have undergone changes with the conversion of pasture land to berry-crop production. MERRIAM'S SHREW and PREBLE'S SHREW are associated with shrub-steppe habitats in the southern Interior, particularly the Southern Okanogan Basin ecosection, a region where there has been loss of natural habitat from development and agriculture.

Island faunas are particularly sensitive to ecological disturbance. Roughly one-third of the shrew and mole species listed as At Risk by the IUCN are endemic to islands. The only island population of shrews listed by the province is the Vancouver Island subspecies of the WESTERN WATER SHREW, which is restricted to Vancouver Island. Associated with wetlands and streams, it may be sensitive to habitat degradation. Four subspecies of PACIFIC SHREW are endemic to small BC islands, and although they are not listed by the province, they warrant more research to determine their taxonomic validity and population status.

The effect of climate change on BC shrew and mole distributions and populations is unknown. A few species such as PREBLE'S SHREW and MERRIAM'S SHREW could undergo range expansions if grasslands expand into forested habitats with global warming. Of the listed species, the TUNDRA SHREW is the most likely to be negatively impacted by climate change. Confined to Arctic tundra–like habitat by the Haines Highway in extreme northwestern BC and isolated far beyond rescue from the nearest known populations in the Yukon or Alaska, vegetational changes could result in the extinction of the BC population.

References

Amori et al. (2011); BC Ministry of Environment (2014); Boyle et al. (1997); Carey and Harrington (2001); Cook, Dawson and MacDonald (2006); Environment Canada (2014, 2016); Kennerley et al. (2021); Nagorsen (2016); Sullivan and Sullivan (1982).

Studying Shrews and Moles

Small and secretive, shrews are rarely seen in the wild. The observer usually gets no more than a quick glimpse of a shrew as it scurries through the forest litter or across the surface of snow. Although molehills are conspicuous, their occupants rarely venture above ground. Another frustration for the naturalist or biologist trying to observe these mammals in the wild is the difficulty of reliably identifying species, even for live animals held in the hand (see Identifying Shrews and Moles, page 47). As a precautionary note on identification, Jordan Ryckman, in her study of a shrew community in the Cascade Mountains of Washington, found that genetic samples (tail clips) taken from her live captures revealed a significant percentage of misidentification for initial field identifications based on morphology. **Be aware that trapping or handling of live shrews or moles in BC requires permits.**

Information on wild shrews is derived largely from trapping studies. For shrews, the Museum Special model of snap trap has been the mainstay for many small-mammal studies and museum collecting in the past. It resembles the hardware-store mousetrap but is less likely to damage the skull, which is an important diagnostic feature. However, for most ecological studies, traps for capture and release are preferred. Live capture allows the researcher to release animals alive in order to obtain information on movements, population numbers and demographics.

Pitfall traps are effective devices for capturing shrews and may be more effective than conventional live traps for some shrew species. The few captures of PREBLE'S SHREW and MERRIAM'S SHREW from BC, for example, were all taken in pitfall traps. Plastic dairy containers (20 centimetres high with a 14-centimetre diameter), small plastic pails or two-litre soda bottles with the tops removed will function as pitfalls. They are set with their rims flush to the surface of the soil. No bait is required, and the animals simply tumble into the trap. Multiple captures in a single pitfall trap are common. The use of drift fences (constructed from 25- to 30-centimetre-wide plastic or aluminum sheeting) to connect traps will direct small-mammal movements and improve trapping efficiency. The disadvantage of pitfall traps is that they are difficult to set in rocky terrain. To avoid pitfall traps from filling with water in heavy rains, position a rain cover above the trap. By providing rain covers, dry nesting material and a food source such as mealworms or blowfly pupae, and checking the trap several times in a 24-hour period, pitfall traps can be used as live-capture traps.

Other trap designs for capturing live shrews are the Sherman and Longworth traps. Both have a door that is closed by a treadle device. The Sherman trap comes in various sizes; one model is a folding design that is easy to transport. Its disadvantages are that bait and nesting material can interfere with the trap release, the treadle is less sensitive than in Longworth traps, and removing live shrews for handling can be tricky. The Longworth trap consists of a tunnel section and a larger nest box that clip together. The door of the tunnel is triggered by a treadle at the rear. The Longworth is bulky to transport and expensive, but its nest box seems to ensure a high survival rate.

A successful bait for live traps is a mixture of peanut butter and rolled oats or invertebrate larvae. A suitable nesting material is coarse cotton batting, which provides good insulation. Live traps have to be checked frequently (four to six times a day) to ensure that captured animals do not starve. There is considerable debate about the relative effectiveness of Sherman, Longworth or pitfall traps for capturing shrews. Which trap you select will depend on the objectives of the inventory, and in some situations, using several types of live traps may be advantageous. A specialized trap for water shrews is a minnow or funnel trap set in shallow water, modified with a platform inside set above water height.

European mammalogists have developed several non-invasive methods for sampling shrews with PVC tubes baited with blowfly pupae. One approach used styrene strips coated with a glue specialized for collecting mammalian hair set at the entrance of the tube to remove hairs from the back of any shrew that enters the tube. Some species, particularly EURASIAN WATER SHREWS, can be reliably identified from their hair morphology. Another approach was to collect shrew fecal droppings left in baited tubes. Droppings of EURASIAN WATER SHREWS were identifiable by the presence of aquatic invertebrate remains. DNA sequencing could also be applied to identify either captured hairs or shrew droppings. The advantage of bait tubes is that shrews are free to exit the tubes. Therefore, they don't require frequent monitoring to prevent fatalities from starvation.

Most traps used for shrews will capture AMERICAN SHREW-MOLES, although pitfall traps seem be most effective. Capturing the fossorial COAST MOLE and TOWNSEND'S MOLE in their underground tunnels is challenging. Moles can be dug out of their breeding nests, but this is very time consuming and moles may be injured. A modified kill trap has been used successfully by several researchers for catching live TOWNSEND'S MOLES. A spring-activated spike trap attached to a metal plate is mounted on a rectangular box (10 by 46 centimetres) that is open at both ends. When activated, the trigger device closes a set of one-centimetre-diameter rods at each end attached to a plate above the trap box. The rods drop to close both ends of the box. An active mole tunnel has to be carefully excavated to set the trap in the tunnel. The Friesian or Dutch tunnel trap, which is based on a similar principle, has been used for centuries to capture the EUROPEAN MOLE. It consists of a wooden base with a mole tunnel bored through it; the trigger releases wooden doors that close off the ends of the passages after a mole enters the trap.

Recapture data give a rather limited picture of a mammal's day-to-day activity or movements, and continuous monitoring is more informative. One ingenious

A funnel minnow trap modified for the live capture of water shrews.

A Longworth live trap.

A folding-type Sherman live trap set by a log.

Live shrew in a pitfall trap.

approach was to tag moles with radioactive chemicals and then track their underground movements with a Geiger counter. Using this method, Valentin Schaefer followed the movements of a COAST MOLE for several months. More recently, miniature radio-transmitters have been glued on the tails of EUROPEAN MOLES and the backs of larger shrews such as EURASIAN WATER SHREW and COMMON SHREW to track individual movements.

Trapping or live capture is not the only tool available for inventorying these animals. Shrew or mole skulls and skeletons recovered in owl pellets can be a source of distributional records. A study in Alberta based on rodent and shrew remains found in owl pellets resulted in new provincial distributional records for VAGRANT SHREW and PRAIRIE SHREW. An indirect sampling method for detecting the presence of COAST MOLE and TOWNSEND'S MOLE that doesn't require capturing animals is their molehills. In the small region where they co-occur, molehills of the two species may be distinguishable by size.

Despite the requirements of permits and specialized traps for biologists, students or wildlife consultants studying shrews and moles, they aren't necessary for naturalists and citizen scientists. We encourage you to post photographs of molehills or moles and shrews found dead to iNaturalist (inaturalist.org). The species accounts and identifications sections will describe which diagnostic features to photograph for reliable identification. Dead shrews or moles can be submitted to a natural history museum (such as the Royal BC Museum or Beaty Biodiversity Museum at UBC) for their preservation as voucher specimens, confirmed identification and possible genetic sampling. Contact museum curatorial or collection staff for advice and assistance with donating specimens.

References

Anthony et al. (2005); Churchfield, Barber and Quinn (2000); Craig (1995); Engley and Norton (2001); Moore (1940); Pocock and Jennings (2006); Rychlik, Ruczynski and Borowski (2010); Ryckman (2020); Schaefer (1982); Sheehan and Galindo-Leal (1997); Stromgren and Sullivan (2014).

Vanessa Craig weighing a live shrew.

Identifying Shrews and Moles

One of the most challenging aspects of studying shrews and moles is how to identify them. A few distinctive species such as PACIFIC WATER SHREW, WESTERN WATER SHREW, ARCTIC SHREW or AMERICAN SHREW-MOLE can be reliably identified from observation or photographs. Others may be identifiable from diagnostic external traits or measurements taken on live captures or dead animals held in the hand, but most shrews will require an examination of cranial/dental traits or measurements to confirm their identification. In some cases a genetic sample for DNA sequencing is required. As a starting point, before you apply the identification keys, we recommend you scan the range maps in the species accounts. This will narrow down the list of potential species for your observation or capture.

Two separate identification keys are included: one for live animals in the hand and another for skull dental traits. They include the 17 species known to occur in BC; for identification traits of the two hypothetical species that could occur in the province, consult their species accounts. The keys are dichotomous, with the diagnostic characteristics arranged into couplets; each couplet offers two mutually exclusive choices (labelled **a** or **b**). To identify your animal, begin with couplet number one and select **a** or **b**. The couplets will either give you a species name or direct you to another couplet in the key. By working through the steps in the key, you will eventually arrive at an identification. We tried to avoid subjective traits (e.g., "slightly darker than" or "slightly larger than") and instead emphasized unique characteristics, size measurements, dental traits and absolute colour differences. To simplify the key, we limited the diagnostic criteria in each couplet to one or two characteristics. Once you obtain an identification, consult the appropriate species account to determine if it is consistent with your determination from the key.

Identification Key to Live Animals in the Hand

This key is intended for identifying restrained live captures of shrews or moles. You will require a small millimetre rule for body measurements and a scale for measuring body mass. Note that some selections lead to a list of multiple species. They are species that cannot be reliably discriminated alive in the hand and will require examination of their skull/dental traits or a genetic sample to confirm their identification. Consult appendix 2 for a summary of external traits of each BC shrew species.

Three external measurements are used in this key; their values for each species are given in the species accounts.

Measurements Used in the Key

- **Total length.** Lay the animal on a millimetre ruler, gently holding the head down and extending the tail. The measurement is from the tip of the nose to the end of the last tail vertebra, which can be felt through the skin with a fingernail.
- **Tail vertebrae length.** Raise the tail at 90 degrees to the back and measure from the back to the tip of the last tail vertebra.
- **Hind foot length.** This is taken by flexing the foot so that it is at right angles to the leg and measuring from the back of the heel to the tip of the longest claw.

Identification Key

1

1a Front feet with broad palms, claws stout and long: **Go to 2**

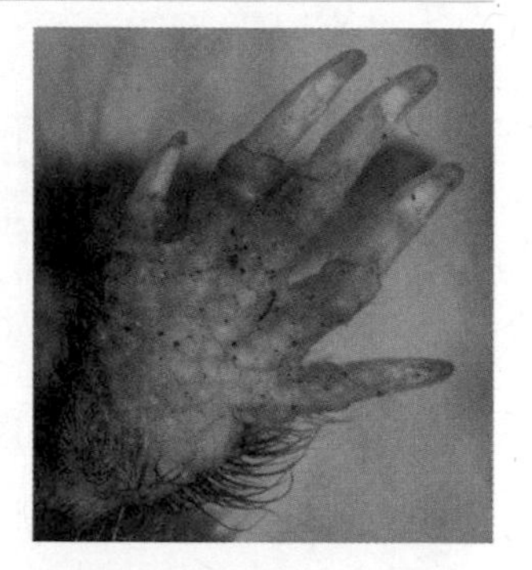

1b Palms of front feet not broad, claws not long and stout: **Go to 4**

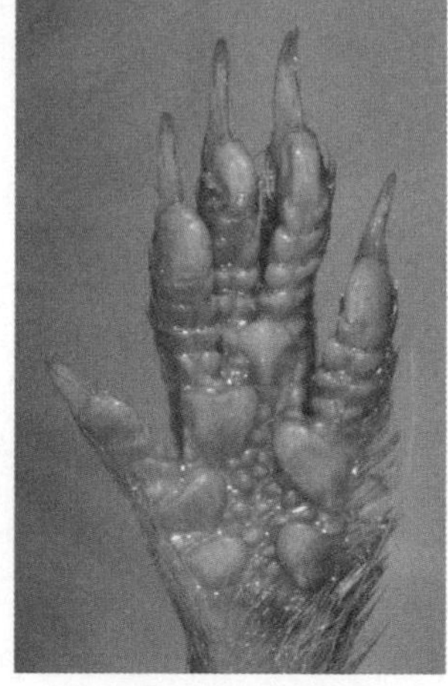

2 (↑1)	**2a**	Tail length more than 25% of total length, width of forefoot palm less than its length, body mass less than 20.0 grams: AMERICAN SHREW-MOLE (p. 71)
	2b	Tail length less than 25% of total length, width of forefoot palm equal or greater than its length, body mass greater than 20.0 grams: **Go to 3**
3 (↑2)	**3a**	Total length greater than 175 mm, hind foot length greater than 24 mm: TOWNSEND'S MOLE: (p. 87)
	3b	Total length usually less than 175 mm, hind foot length less than 24 mm: COAST MOLE (p. 79)
4 (↑1)	**4a**	Hind foot length usually greater than 18 mm with a fringe of stiff hairs: **Go to 5**

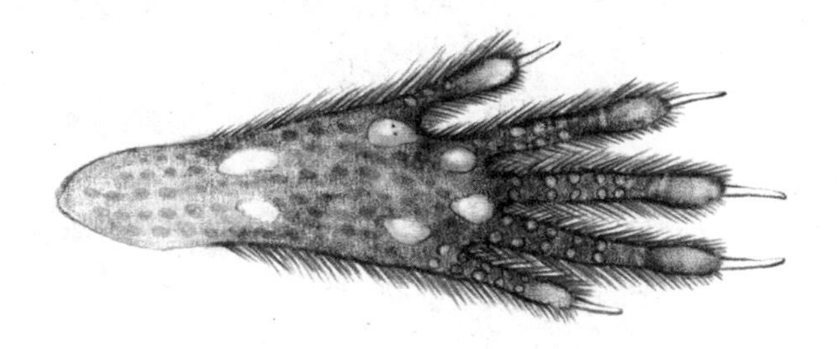

	4b	Hind foot length usually less than 18 mm and lacking a fringe of stiff hairs: **Go to 6**

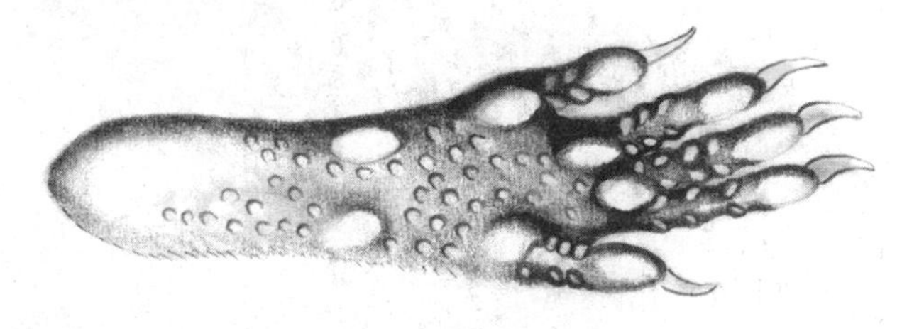

5 (↿4) **5a** Dorsal fur black or dark grey, feet and belly grey or silver grey, tail distinctly bicoloured with a paler underside, distal end of tail has a ventral keel: AMERICAN WATER SHREW (p. 151) or WESTERN WATER SHREW (p. 129)

In northeastern BC where their ranges may overlap, use the Identification Key to Cranial/Dental Traits (p. 52) or take genetic samples to distinguish

5b Dorsal fur dark brown, feet and belly dark brown, tail dark brown and not bicoloured, distal end of tail lacks a ventral keel: PACIFIC WATER SHREW (p. 101)

6 (↿4) **6a** Pelage appears tricoloured or saddlebacked, fur on the back contrasts sharply to paler sides and belly: **Go to 7**

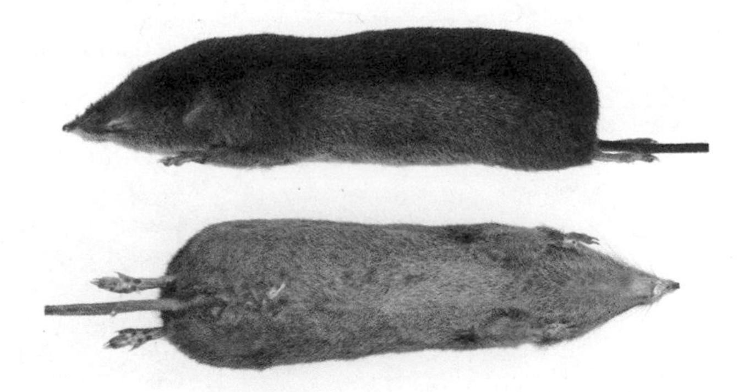

6b Pelage does not appear tricoloured or saddlebacked, fur on the back does not contrast sharply to fur on the sides: **Go to 8**

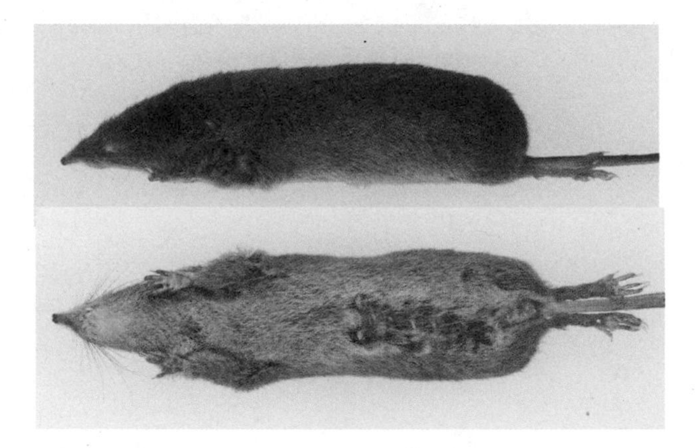

7 (↰6) **7a** Fur on the back is blackish or dark brown and side and belly fur is a lighter brown, tail length usually greater than 37 mm, range is northeastern BC east of the Rocky Mountains: ARCTIC SHREW (p. 95)

7b Fur on the back is brown and side and belly fur is paler greyish brown, tail length usually less than 37 mm, range is extreme northwestern BC: TUNDRA SHREW (p. 177)

8 (↰6) **8a** Tail length usually less than 34 mm and less than 40% of total length: WESTERN PYGMY SHREW (p. 117)

8b Tail length usually greater than 34 mm and more than 40% of total length: **Go to 9**

9 (↰8) **9a** Dorsal pelage dark grey, tail distinctly bicoloured with the dorsal side dark grey and a white ventral side contrasting with a dark belly: TROWBRIDGE'S SHREW (p. 171)

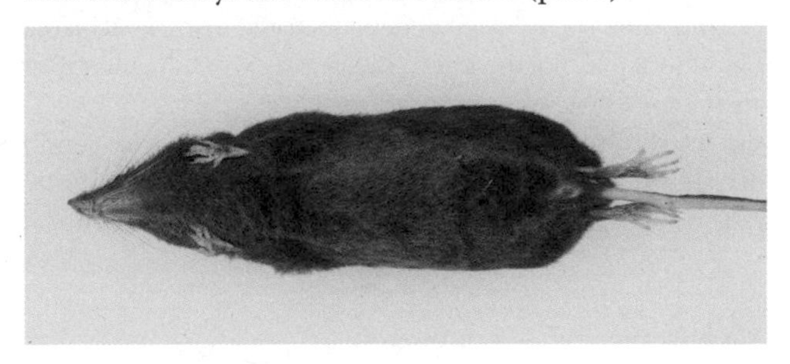

9b Dorsal pelage brown to grey, tail distinctly to indistinctly bicoloured with the dorsal side brown to grey and a variable paler ventral side not contrasting with a dark belly (see images for 6b on p. 50): CINEREUS SHREW (p. 109), DUSKY SHREW (p. 137), MERRIAM'S SHREW (p. 123), PACIFIC SHREW (p. 143), OLYMPIC SHREW (p. 165), PREBLE'S SHREW (p. 159) or VAGRANT SHREW (p. 183).

These cannot be discriminated from external morphology in the hand. Use the Identification Key to Cranial/Dental Traits (p. 52) or genetic samples for identification.

Identification Key to Cranial/Dental Traits

This key relies on cranial or dental characteristics. It is designed primarily to identify cleaned skulls from museum specimens or skulls found in raptor pellets; however, for dead whole animals it is possible to push back the lips to reveal the incisors and unicuspid teeth for examination. You will require a 15× hand lens or better yet a dissecting microscope to examine teeth.

Although more reliable than the identification key to whole animals, this key for cranial/dental traits has its limitations. Because of their fragility, skulls are occasionally broken. Skull length, a diagnostic trait in the key, cannot be measured on broken skulls. Old individuals with badly worn teeth also can be difficult to identify because the pigmentation patterns on the upper and lower incisors and the relative size of the upper unicuspid teeth may be obscured. In old individuals, the pigmented region may be completely worn away, along with the accessory tines on the upper first incisors.

Some shrew species are difficult to distinguish based on one or two diagnostic characteristics in a dichotomous key. We used diagnostic character traits that are typical of most individuals, but a trait may vary with a few individuals of a species showing atypical character states. Consult appendix 3 for a summary of multiple cranial/dental traits for each of BC's shrew species. Dead animals can be submitted to a natural history museum for identification confirmation.

Measurements Used in the Key

The following measurements are used in the key. Measuring skulls and teeth can be done with calipers (needlepoint calipers are best) or an ocular micrometer set in the eyepiece of a dissecting microscope.

Skull length. For moles, skull length is the greatest overall length of the skull; for shrews, it is taken from the anterior edge of the premaxilla between the front two incisors to the greatest posterior point of the skull.

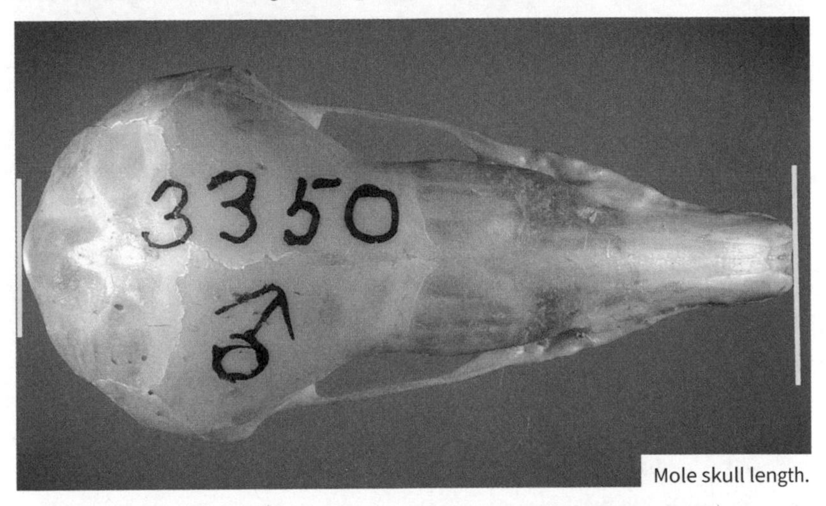

Mole skull length.

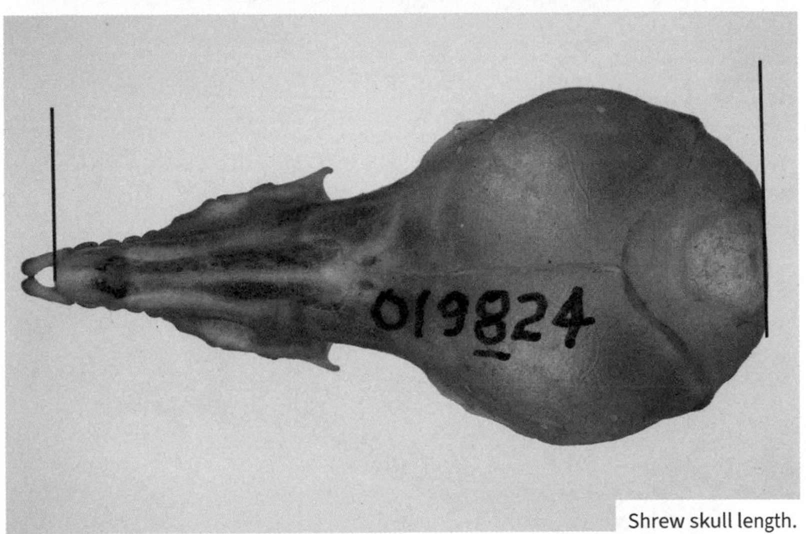

Shrew skull length.

Palatal length. Measured from the most anterior point on the palate between the most anterior incisors and the most posterior edge of the palatal shelf.

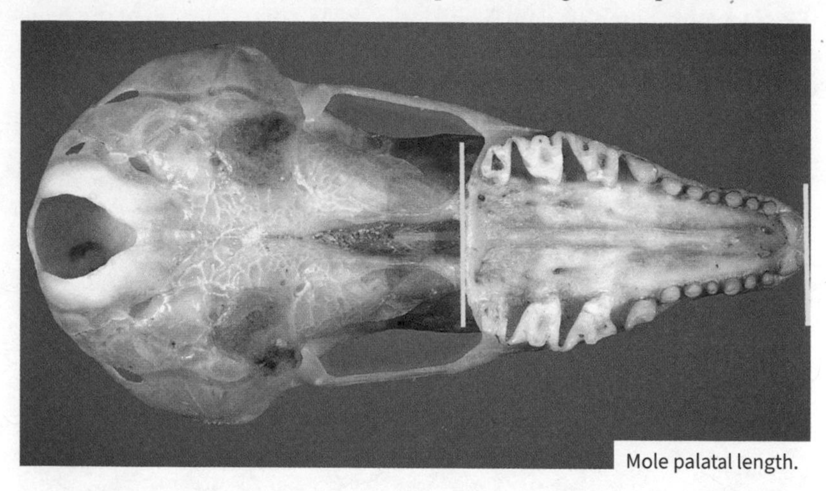

Mole palatal length.

Width of first upper unicuspid. Greatest width across first upper unicuspid behind the incisor. This measurement is best taken with an ocular micrometer in a dissecting microscope.

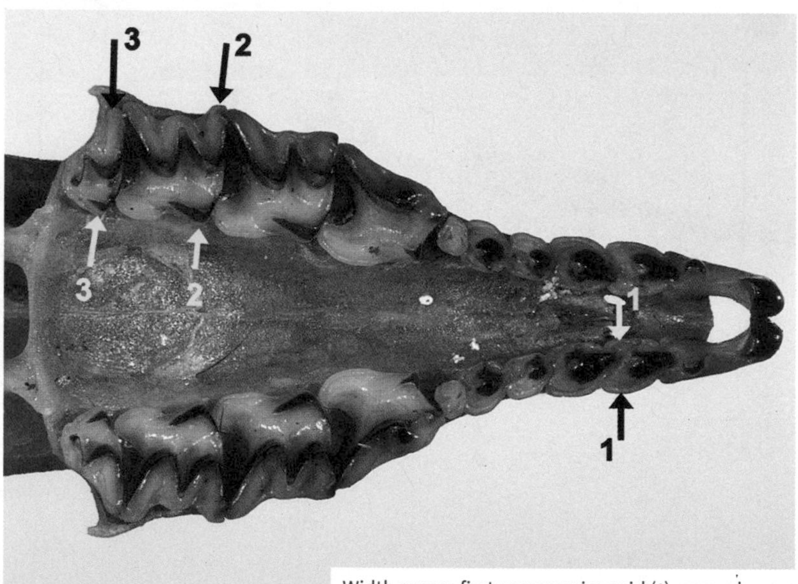

Width across first upper unicuspid (1), second upper molar (2) and third upper molar (3) of a shrew.

Width across second upper molars. A measurement across the greatest width.

Width across third upper molars. A measurement across the greatest width.

Height of coronoid process. Shortest distance from the tip of the coronoid process to the ventral edge of the dentary.

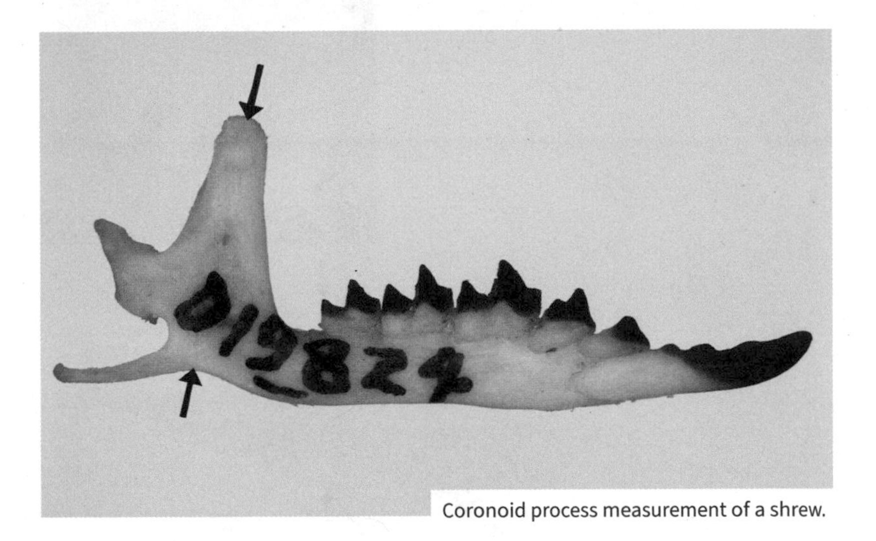

Coronoid process measurement of a shrew.

Identification Key

1

1a Skull with thin zygomatic arches, first upper incisor not enlarged and lacking two elongated cusps: **Go to 2**

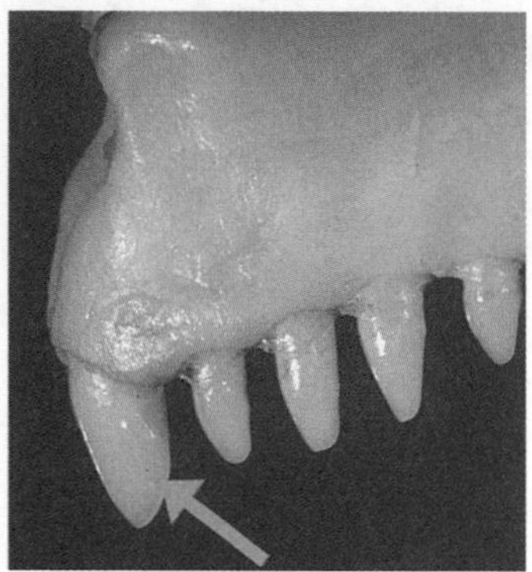

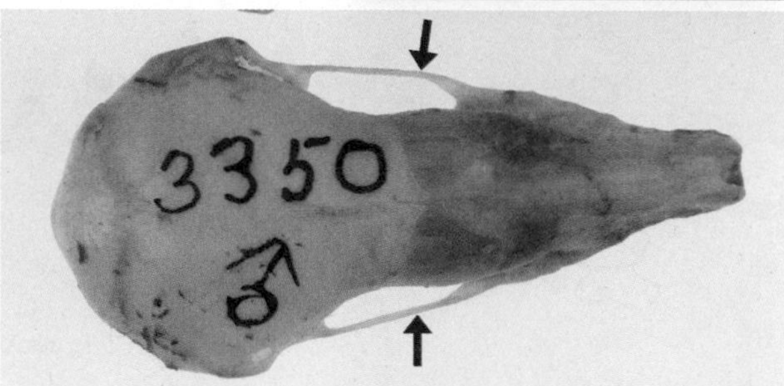

1b Skull lacking zygomatic arches, first upper incisor enlarged with two elongated cusps: **Go to 4**

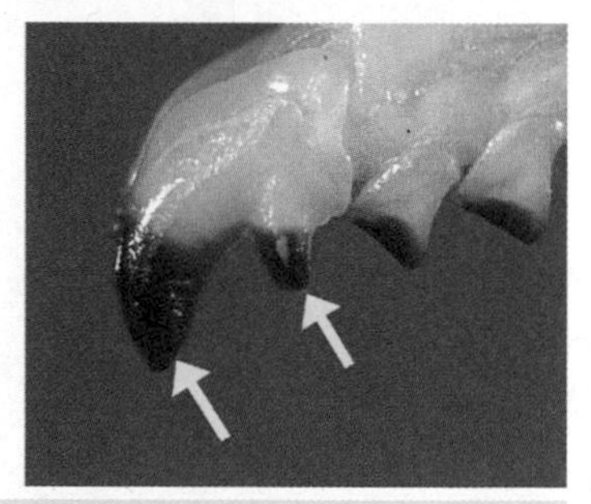

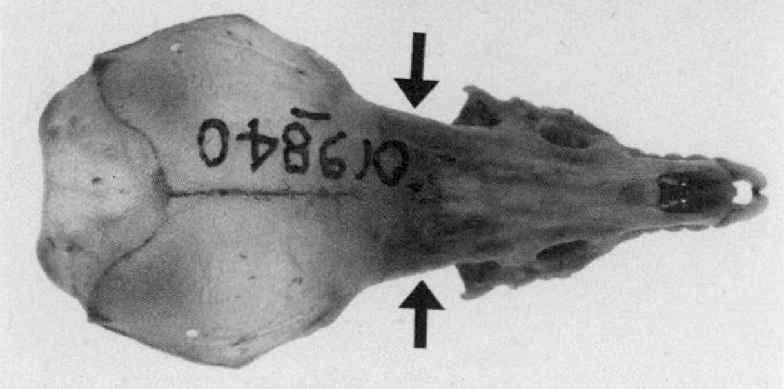

2 (↑1) **2a** Skull length less than 25.0 mm, palatal length less than 13.0 mm; skull with two upper incisors and three upper premolars on each side: AMERICAN SHREW-MOLE (p. 71)

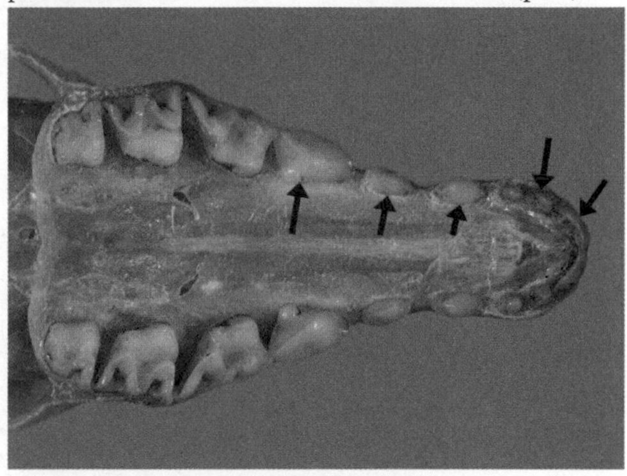

2b Skull length more than 25.0 mm, palatal length greater than 13.0 mm; skull with three upper incisors and four upper premolars on each side: **Go to 3**

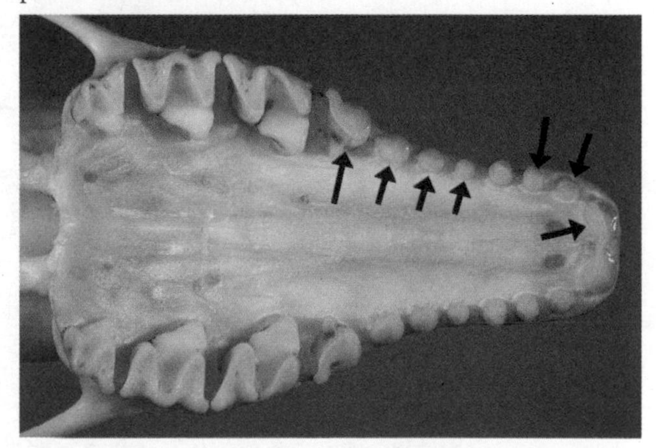

3 (↑2) **3a** Skull length greater than 37.0 mm, palatal length greater than 16.0 mm: TOWNSEND'S MOLE (p. 87)

3b Skull length less than 37.0 mm, palatal length less than 16.0 mm: COAST MOLE (p. 79)

4 (↰1) **4a** Third upper unicuspid a flattened disc (ventral view), five unicuspids but only three unicuspids easily visible from the side, face of upper incisor has a deep groove beside the long medial tine: WESTERN PYGMY SHREW (p. 117)

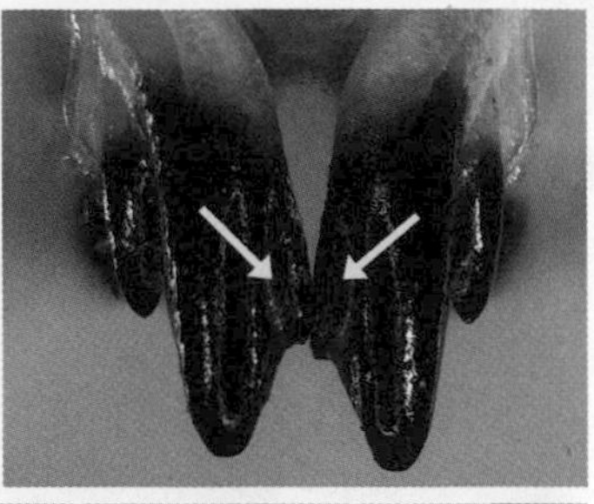

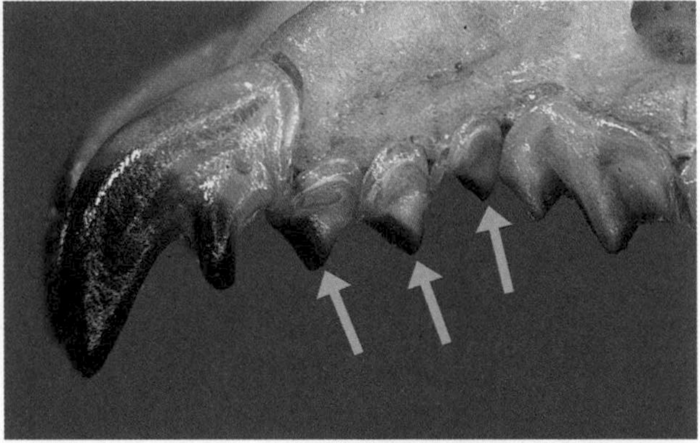

4b Third upper unicuspid not minute, four or five unicuspids visible in side view, face of upper incisor not grooved: **Go to 5**

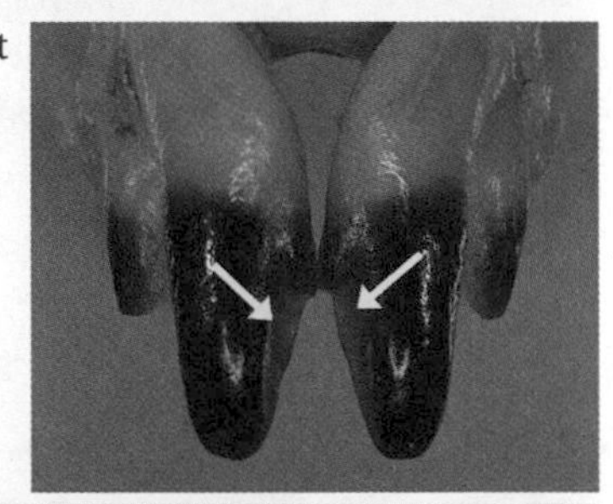

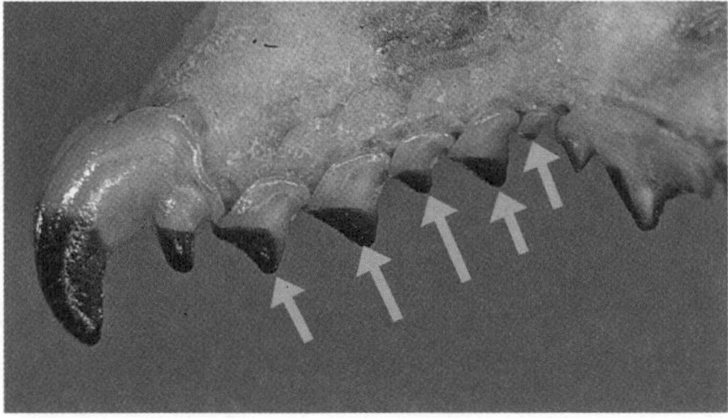

5 (↰4) **5a** Lower incisor with very shallow indentations between the three denticles, the third denticle almost imperceptible, medial tine absent on upper incisor, interior edge of the upper incisors forms a continuous curve: MERRIAM'S SHREW (p. 123)

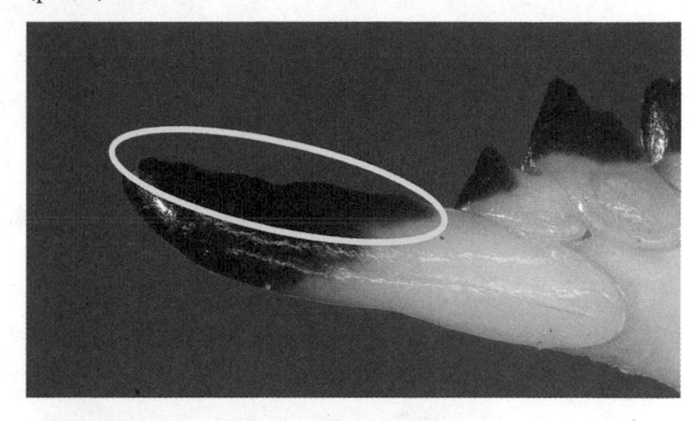

5b Lower incisor with deep indentations between the three denticles, all denticles clearly visible, medial tines present on upper incisor, interior edge of the upper incisors forms a flattened area where the incisors come together: **Go to 6**

Medial tines are minute or absent in some individual OLYMPIC SHREWS, although there is a flattened area where the two incisors come together. This species's range is outside that of MERRIAM'S SHREW.

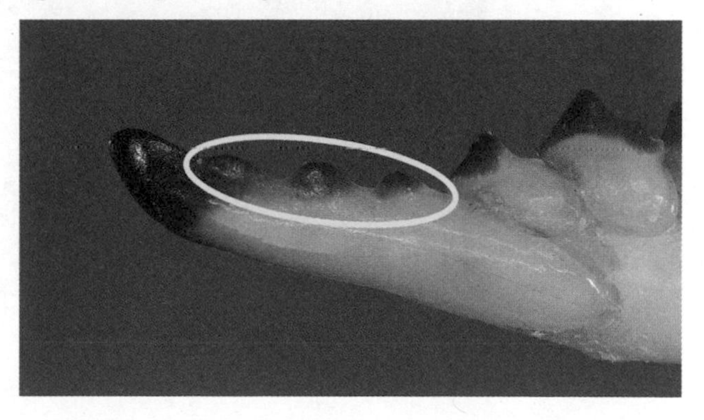

6 (↑5) **6a** First upper molar usually has a spot of pigment on the hypocone, pigment on lingual side of lower incisor only extends to first denticle: **Go to 7**

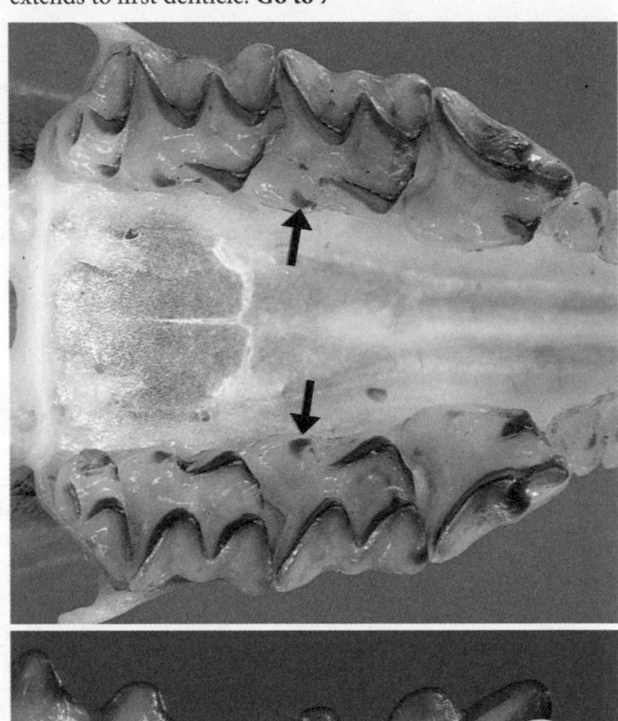

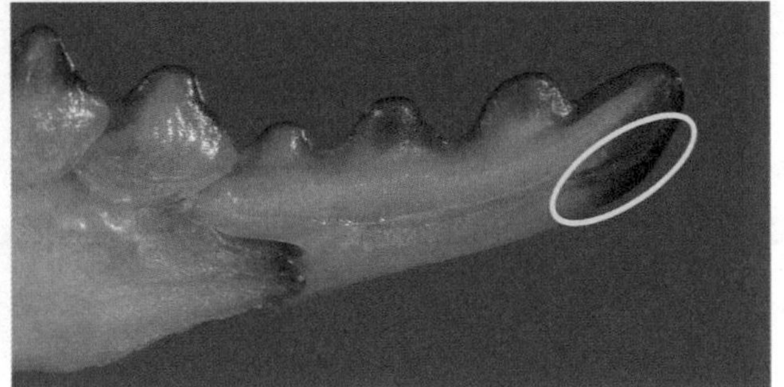

6b First upper molar lacking pigment on the hypocone, pigment on lingual side of lower incisor usually extends beyond first denticle: **Go to 8**

The first upper molar hypocone is rarely pigmented in CINEREUS SHREW; the pigment on the lingual side of the lower incisor extends to the first denticle in OLYMPIC SHREW.

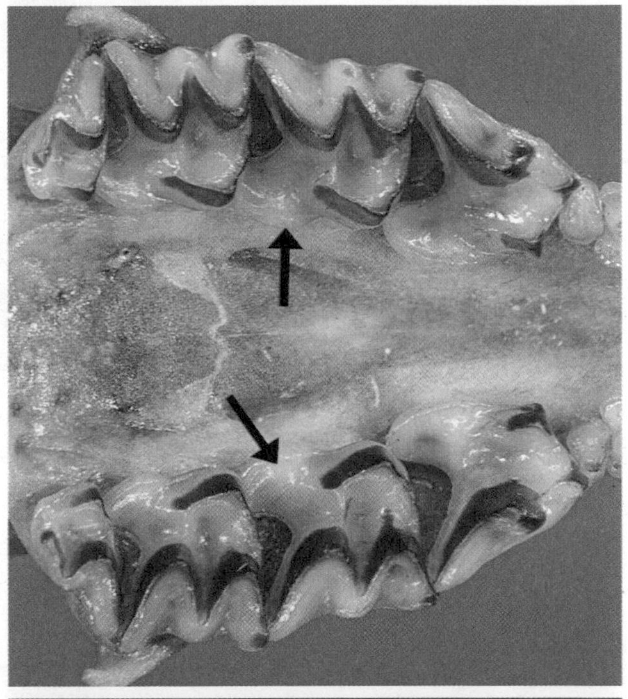

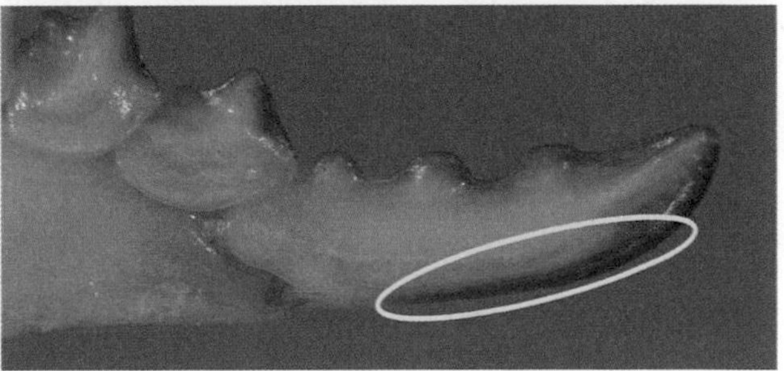

7 (↑6) **7a** Edge of pigment on anterior face of upper incisor is above the pigment on lateral cusp: TUNDRA SHREW (p. 177)

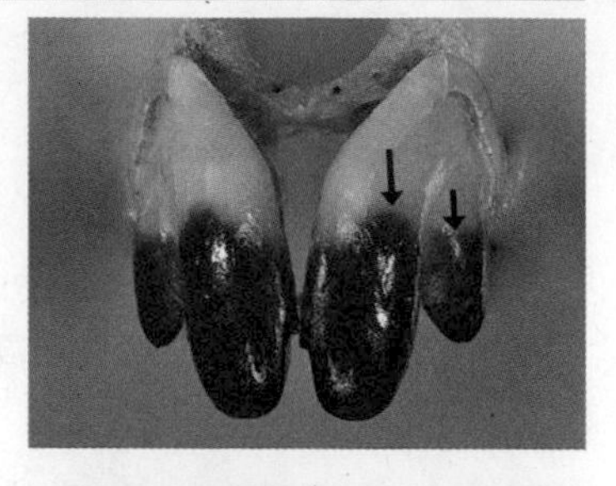

7b Edge of pigment on anterior face of upper incisor is at or below the pigment on lateral cusp: ARCTIC SHREW (p. 95)

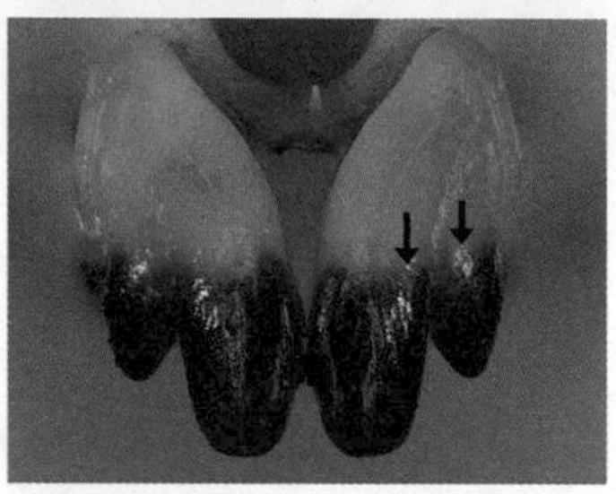

8 (↑6) **8a** Skull length less than 14.5 mm: PREBLE'S SHREW (p. 159)

8b Skull length greater than 14.5 mm: **Go to 9**

9 (↑8) **9a** Skull length greater than 18.7 mm: **Go to 10**

9b Skull length less than 18.7 mm: **Go to 12**

10 (↑9) **10a** Rostrum down-curved in side profile: PACIFIC WATER SHREW (p. 101)

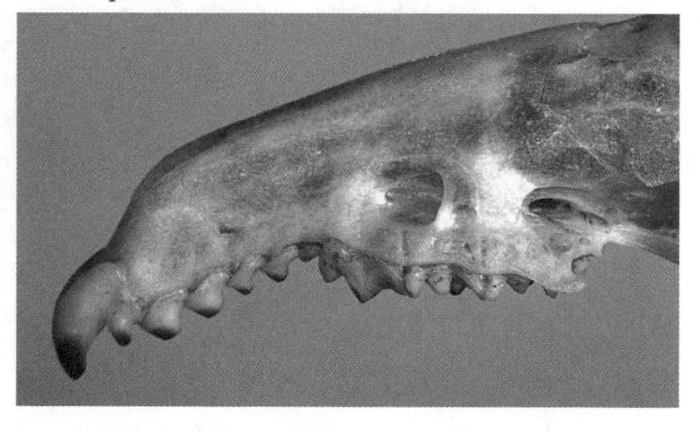

10b Rostrum not down-curved in side profile: **Go to 11**

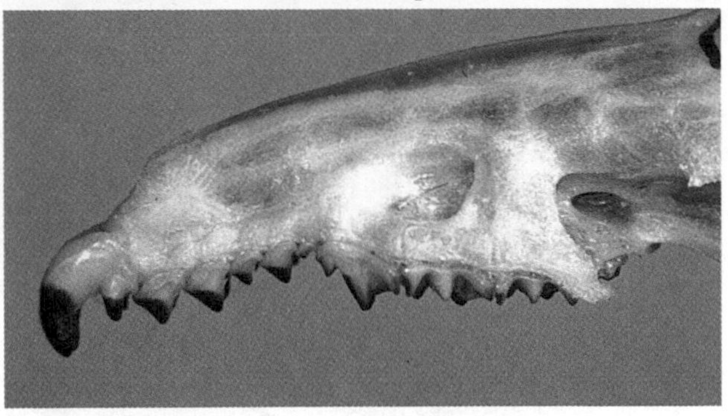

11 (↑10) **11a** Width of first upper unicuspid less than 0.76 mm, height of coronoid process on dentary less than 4.6 mm: WESTERN WATER SHREW (p. 129)

11b Width of first upper unicuspid greater than 0.76 mm, height of coronoid process on dentary greater than 4.6 mm: AMERICAN WATER SHREW (p. 151)

12 (↑9) **12a** Width across third upper molars 4.1 mm or less: **Go to 13**

12b Width across third upper molars 4.1 mm or greater: **Go to 15**

If the width is exactly 4.1 mm, refer to appendix 3 and the species accounts.

13 (↑12) **13a** Width of first upper unicuspid less than 0.56 mm: OLYMPIC SHREW (p. 165)

13b Width of first upper unicuspid greater than 0.56 mm: **Go to 14**

14 (↳13) **14a** Third upper unicuspid shorter than the fourth (rarely equal) (see top picture), medial tine on upper incisor separated from main pigment by a pale gap (see bottom picture): VAGRANT SHREW (p. 183)

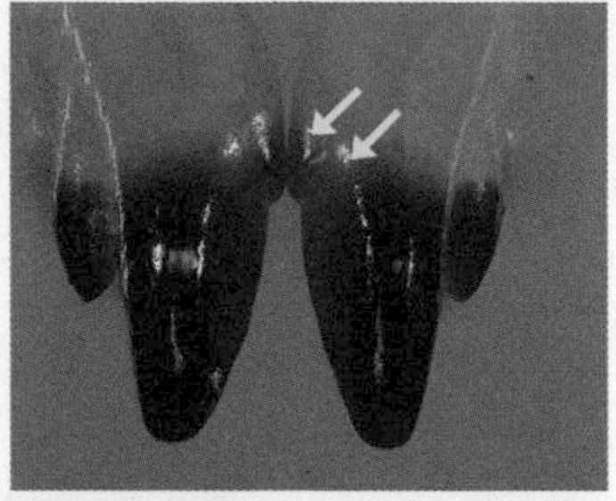

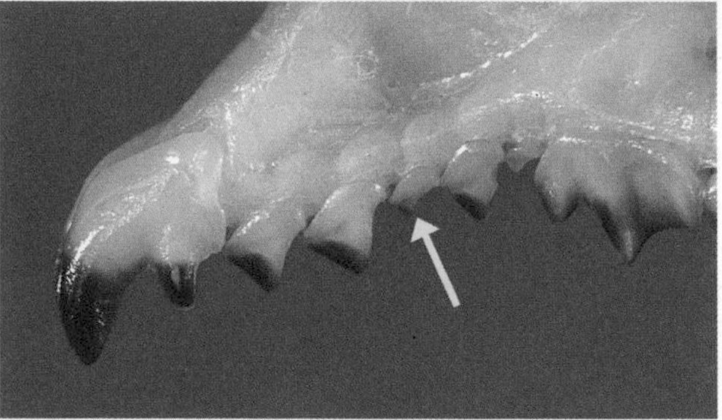

14b Third upper unicuspid taller than or equal to the fourth (rarely shorter) (see top picture), medial tine on upper incisor located in main dark pigment area (see bottom picture): CINEREUS SHREW (p. 109)

A few individuals have a pale edge to the pigment including the medial tine.

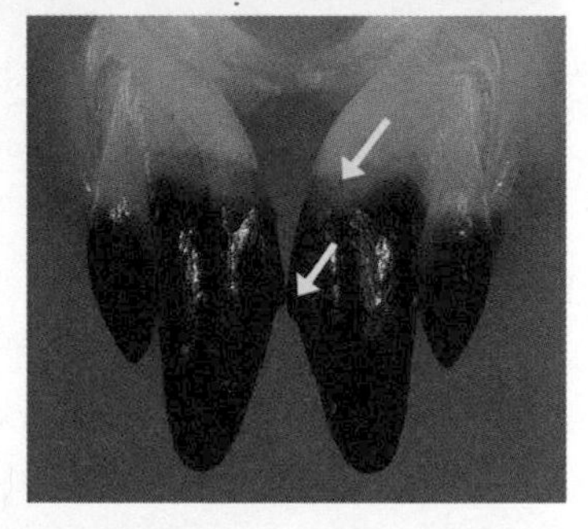

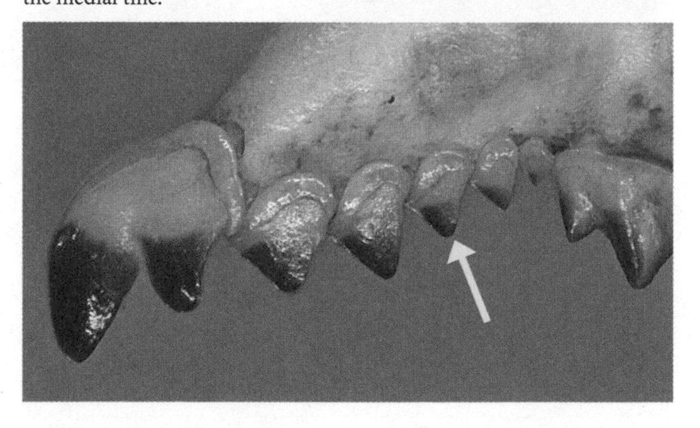

15 (↿12) **15a** Lower portion of first upper unicuspid ridge on the lingual side is unpigmented in specimens with unworn teeth, large postmandibular foramen usually present on both dentaries: TROWBRIDGE'S SHREW (p. 171)

Slight wear on unicuspids creates a noticeable groove between the ridge and the cingulum.

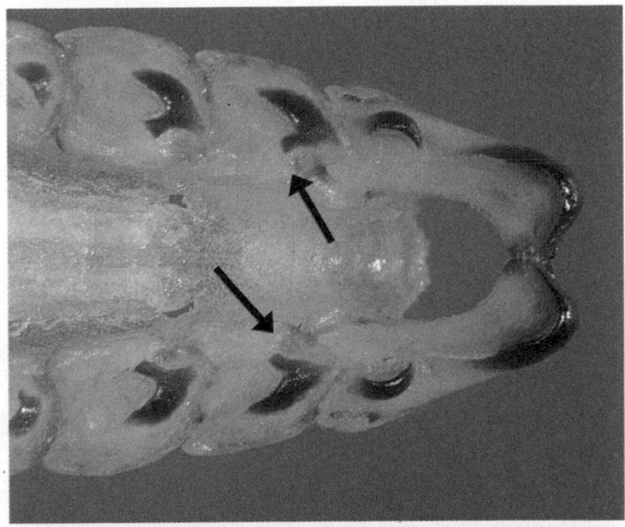

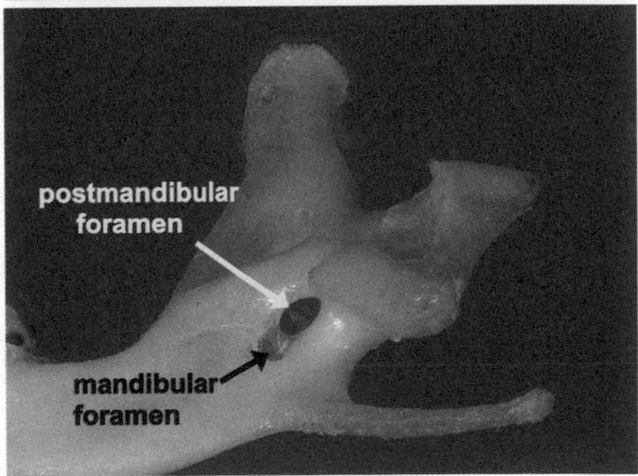

15b First upper unicuspid ridge on the lingual side is pigmented to the cingulum, postmandibular foramen usually absent: DUSKY SHREW (p. 137) or PACIFIC SHREW (p. 143)

The postmandibular foramen is present in about one-third of specimens examined of both species but usually present only on one dentary. These two species cannot be reliably identified from skull characters. Use distribution or genetic information.

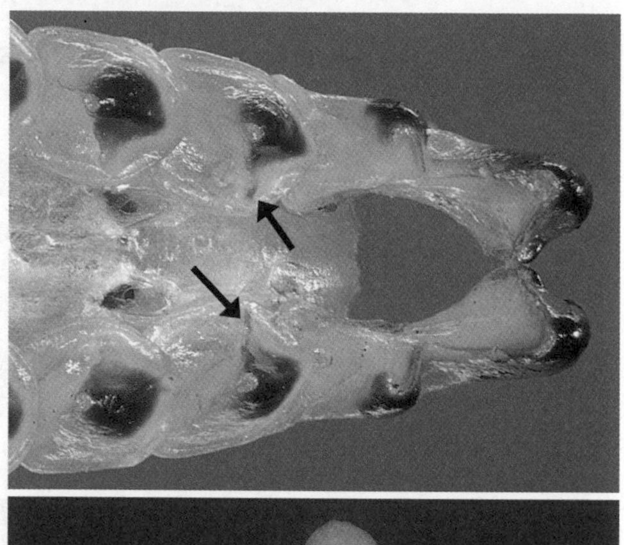

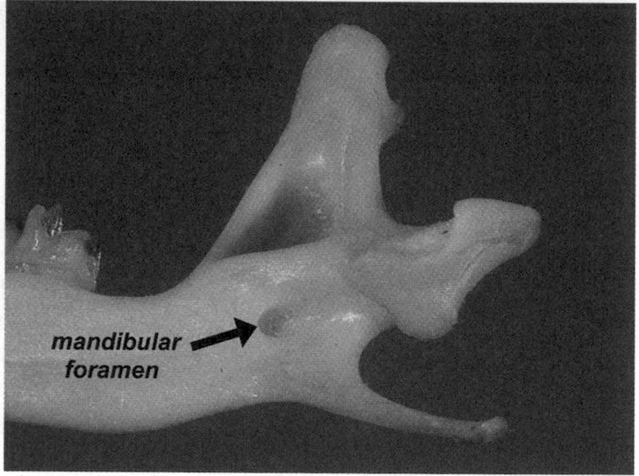

References

Carraway (1987, 1995); Junge and Hoffmann (1981); Nagorsen and Panter (2009); Nagorsen, Panter and Hope (2017); Nagorsen et al. (2001); Van Zyll de Jong (1983a); Verts and Carraway (1998).

Species Accounts

We provide a detailed species account here for each of the 17 species documented to occur in BC. Accounts are arranged in the order of scientific names (see the table of contents). Alternate English common names are given for species with multiple common names. A hypothetical-species section at the end of the species accounts provides a brief synopsis for two shrew species that have no occurrence records from BC but may be found here.

DESCRIPTION provides a concise description of fur colour and other external traits illustrated with photographs taken of a representative museum study skin. For species known from only a few BC specimens (AMERICAN WATER SHREW, TUNDRA SHREW, PREBLE'S SHREW and MERRIAM'S SHREW), our descriptions have been enhanced with information published from neighbouring regions. Cranial/dental traits include the total number of teeth and the dental formula, which describes the number of teeth of each type in the upper and lower jaws on one side of the head. For example, "incisors 2/1, canines 1/1, premolars 3/4, molars 3/3" (the dental formula for the AMERICAN SHREW-MOLE) indicates two upper and one lower incisors, one upper and one lower canine, three upper and four lower premolars, and three upper and three lower molars. The dental formula for all BC shrews is incisors 1/1, unicuspids 5/1, premolars 1/1, molars 3/3. Mammalogists can't agree on which of the unicuspid teeth are incisors, canines or premolars (see Cranial/Dental Traits, page 16).

DISTRIBUTION includes a general description of the species's North American range and a more detailed description of its provincial range. An associated range map shows known occurrence records and the expected range. Species range maps are based on ecosection borders that are consistent with the BC Conservation Data Centre's method for mapping species ranges. Their ecoregion classification system stratifies the province according to geographical units. The three levels (ecoprovinces, ecoregions and ecosections) are progressively more detailed categorizations of the ecosystem characteristics. Categories are based on similar climate, physiography, hydrology, vegetation and wildlife potential. The 10 terrestrial ecoprovinces represent broad geographic areas with consistent climate and terrain. An ecoregion is an area with major physiographic and minor macroclimatic variations. There are 38 pure terrestrial ecoregions in BC. Ecosections are areas with minor physiographic and macroclimatic variations. There are 139 ecosections in BC, ranging from pure marine units to pure terrestrial units.

Confirmed occurrence records (voucher specimens and observations) are plotted as symbols (specimens in blue, observations in yellow) on the map in the species account. Ecosection(s) with occurrence records are represented with dark-grey shading. Note that the entire ecosection is shaded, even if it contains areas where the species may be absent. Some species may have distributions that extend beyond ecosections with documented occurrences. For those species, light-grey shading represents ecosections where the species is expected to occur although there are no documented occurrences. Although somewhat subjective, we identified the expected range using the species's known habitat requirements and the presence of suitable habitat in the ecosection.

For the 14 shrew species, confirmed occurrences consisted almost entirely of museum voucher specimens. Most were examined by the authors to verify their identification. Some had associated genetic data that confirmed their identification. Shrew observational records were mostly limited to the three species of water shrews that can be identified in the hand. We did not use any BC shrew occurrence records found on iNaturalist because their whole-animal photographs were inadequate to make a reliable identification. For the three mole species, confirmed occurrence records consisted of museum voucher specimens and observational records. We used occurrences supported by photographs from iNaturalist of AMERICAN SHREW-MOLE and COAST MOLE, and several peripheral AMERICAN SHREW-MOLE occurrences in our range map are based on iNaturalist observations. Some observational records of COAST MOLE and TOWNSEND'S MOLE were based on molehills rather than a direct observation of the animal.

Species occurrence data were from 21 natural history museums that provided records or their records were accessed through the VertNet (vertnet.org) and Arctos (arctosdb.org) portals. Other records were contributed by citizen scientists, individual researchers, naturalists, wildlife consultants and government sources including BC's Ministry of Environment and Climate Change Strategy (wildlife data and information) and the BC Conservation Data Centre. Natural history museums and individuals that contributed are listed in the acknowledgements (page 197).

MEASUREMENTS includes three linear measurements (see Measurements Used in the Key, page 48) and body mass; linear measurements are given in millimetres and body mass in grams. Our measurements are from BC specimens except for species known from only a few captures in BC where we enhanced our sample with measurements of specimens from neighbouring regions. We give the range of values, the mean, and sample size (n = number of individuals measured). Because mass varies with season and reproductive condition, there is considerable variation associated with the mean for each species. For species with sufficient samples of known sex, we present separate

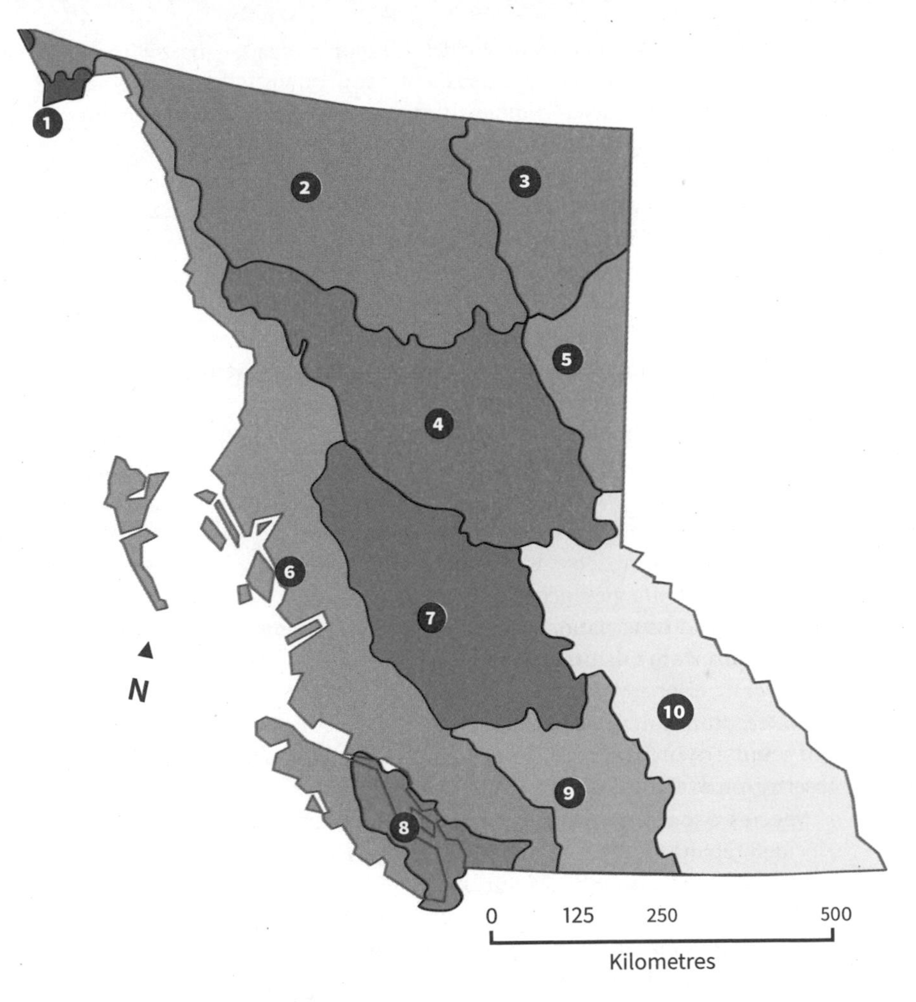

Ecoprovinces of BC

- 1. Southern Alaska Mountains
- 2. Northern Boreal Mountains
- 3. Taiga Plains
- 4. Sub-Boreal Interior
- 5. Boreal Plains
- 6. Coast and Mountains
- 7. Central Interior
- 8. Georgia Depression
- 9. Southern Interior
- 10. Southern Interior Mountains

mean body mass for males and females. Measurements are mostly from data associated with museum specimens and a few published studies. As these measurements were taken by many different individuals, some variation can be expected from differences in measuring technique.

MORPHOLOGICALLY SIMILAR SPECIES is a summary of diagnostic traits for distinguishing the species from morphologically similar species. Note that a number of shrews require skull or dental traits or a genetic sample to confirm their identification.

NATURAL HISTORY includes habitat, elevational range, movements, population estimates, food habits, behaviour and reproduction. Wherever possible, we used information from studies done on BC populations or populations in adjacent areas of western Canada or the western United States. Considerable habitat, elevational and reproductive data were also obtained from data associated with museum specimens collected in the province.

CONSERVATION STATUS lists the conservation status of the species including the provincial listing and national rankings if a species is at risk. Potential and ongoing threats are discussed.

SYSTEMATICS AND TAXONOMY summarizes findings from recent genetic studies and discusses any taxonomic issues. Currently recognized subspecies in BC mostly based on the Neal Woodman's monograph on American Eulipotyphla are listed here. For species represented by two or more subspecies in BC, we give separate linear measurements and body mass for each subspecies based on measurements from BC captures.

REMARKS gives interesting details about the species, recommendations for future study, and the etymology of the scientific species name.

REFERENCES also includes some sources cited in the 1996 edition of this handbook. In keeping with the format of past Royal BC Museum handbooks, we generally do not cite the sources of information directly in the text of the account except for unpublished observations or in a few cases where the source is essential.

American Shrew-mole *Neurotrichus gibbsii*

DESCRIPTION

AMERICAN SHREW-MOLE is a small mole with some shrew-like features, with fur colour ranging from sooty blue-black to nearly black. The fur is directed backward like that of a shrew and lacks the lush, velvety texture characteristic of other BC moles. It has no external ears, and the ear opening is an oval slit; the eyes are minute and hidden in the fur. The nose, long and flattened above and below, is equipped with eight pairs of vibrissae near the base. The nose tip has a fringe of short bristles. The feet and tail are scaly, with front feet longer than they are wide, specialized for digging with long, curved claws. The thick scaly tail is noticeably constricted at the base; it is sparsely covered with short stiff hairs, and a tuft of hairs extends from the tip.

Cranial/dental traits: skull with complete zygomatic arches, auditory bullae incomplete; 36 teeth; incisors 2/1, canines 1/1, premolars 3/4, molars 3/3; first upper incisor enlarged and flattened front and back; upper canine larger than the third upper incisor; lower incisor small and spatula-like directed forward.

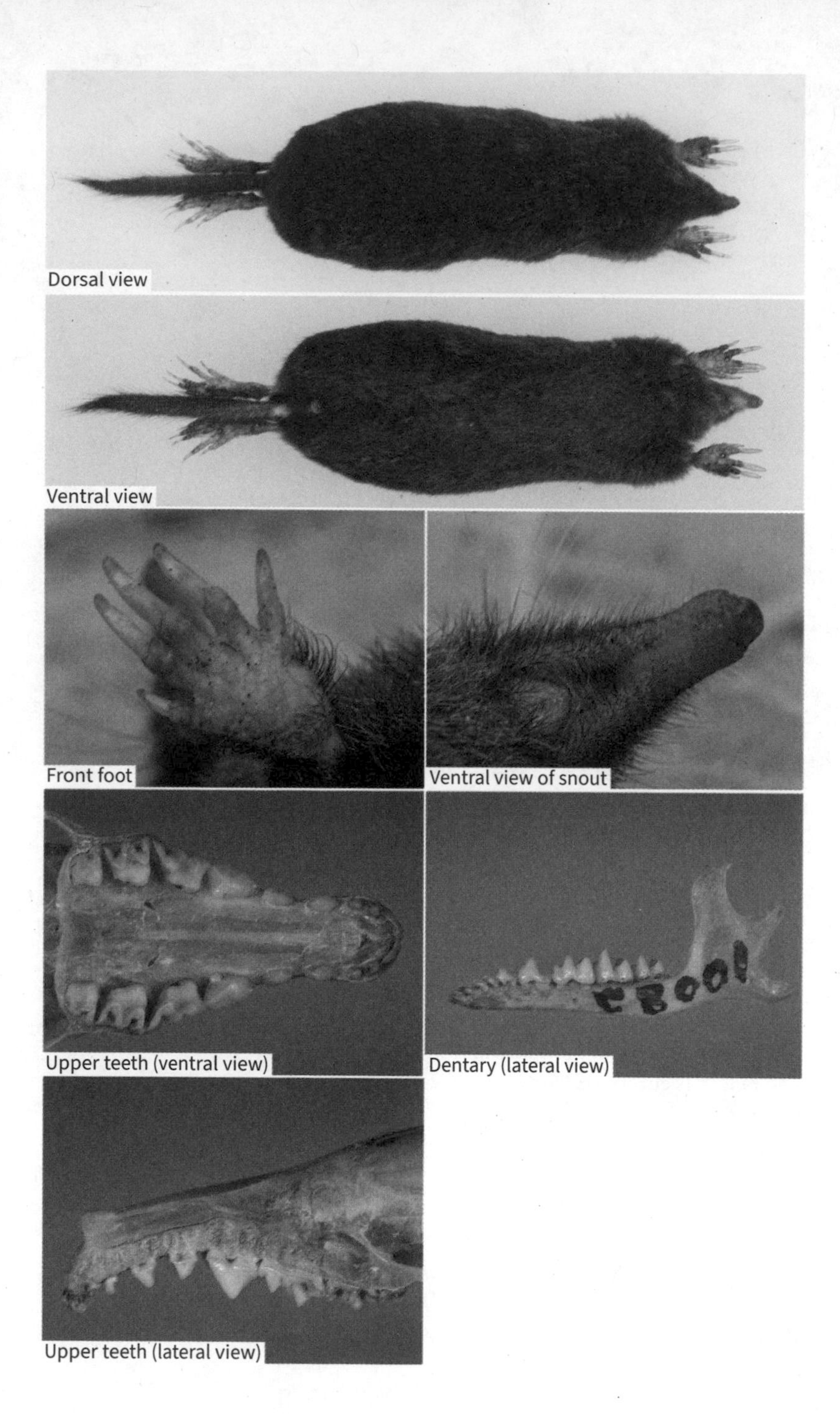
Dorsal view
Ventral view
Front foot
Ventral view of snout
Upper teeth (ventral view)
Dentary (lateral view)
Upper teeth (lateral view)

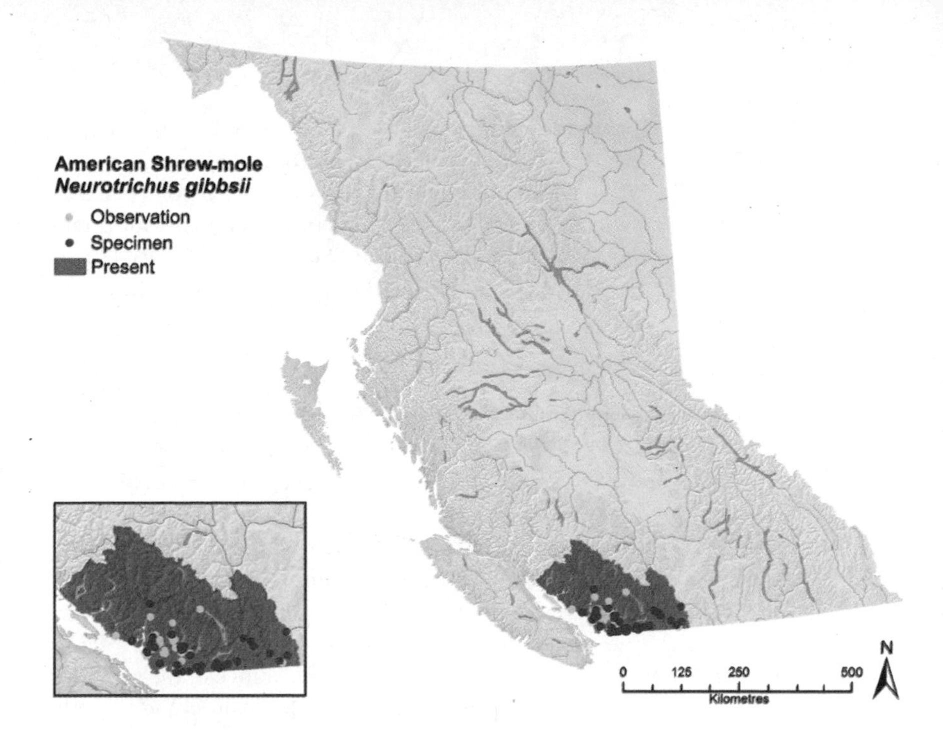

DISTRIBUTION

AMERICAN SHREW-MOLE ranges from California to BC throughout coastal lowlands and coastal mountain ranges (Sierra Nevada and Cascade, Olympic and Coast Mountains). It is widely distributed across southwestern BC, where it inhabits the lower Fraser River valley, Cascade Mountains and North Shore mountains. Northern limits are the Sechelt Peninsula and Squamish; eastern limits are in Manning Provincial Park, although a historical museum specimen captured at Bromley Creek in a wet ravine at 850 metres elevation about five kilometres southwest of Princeton is an outlying eastern record. This occurrence suggests that localized populations inhabit some of the wet valleys on the eastern side of the Cascade Mountains in BC.

MEASUREMENTS

	Mean	Range	Sample size
Total length:	112 mm	98–125 mm	$n = 183$
Tail vertebrae:	37 mm	29–50 mm	$n = 184$
Hind foot:	16 mm	14–19 mm	$n = 184$
Mass—females:	11.3 g	8.3–14.5 g	$n = 8$
Mass—males:	10.9 g	8.0–13.6 g	$n = 22$

MORPHOLOGICALLY SIMILAR SPECIES

AMERICAN SHREW-MOLE is sometimes mistaken for a shrew, but the enlarged front feet with long claws, thick scaly tail, and dentition readily distinguish it from any shrew. In external traits, it differs from COAST MOLE or TOWNSEND'S MOLE by having a smaller body size (a mass less than 20.0 grams), longer tail, and front paws that are longer than they are wide. Skulls of COAST MOLES and TOWNSEND'S MOLES are larger (a skull length greater than 25.0 millimetres) with 44 teeth.

NATURAL HISTORY

AMERICAN SHREW-MOLE is generally associated with moist coniferous and mixed forests with rich deep soils and an extensive ground cover of decaying logs and stumps. Typical vegetation includes a tree cover of WESTERN REDCEDAR, WESTERN HEMLOCK, BIGLEAF MAPLE or RED ALDER, and a shrub layer of SALMONBERRY, DULL OREGON GRAPE and RED ELDERBERRY. Nevertheless, it is not particularly specialized in its habitat requirements and can also be found in riparian areas, lakeshore swamps, the forest edge, SKUNK CABBAGE marshes and damp meadows. Several researchers reported that dark, damp ravine bottoms provide the ideal habitat. Its elevational range in the province extends from sea level to 1,380 metres on Liumchen Mountain in the Cascade Mountains and up to 940 metres in the North Shore mountains.

AMERICAN SHREW-MOLE's association with forests of different ages is not clear; it has been captured in recent clearcuts and second-growth and old-growth forests. Several studies in the DOUGLAS-FIR forests of Oregon and Washington found a strong affinity for moist old-growth forests. However, studies in other parts of the western United States revealed no clear relationship between abundance and forest age. In a major survey of small mammals in the Greater Vancouver watersheds in the southern Coast Mountains, Dale Seip found that AMERICAN SHREW-MOLES were most common in second-growth forest.

More active above ground than COAST MOLE or TOWNSEND'S MOLE, it bends the front claws inward when walking and supports its weight on the backs of them. This mole is surprisingly agile and can move quickly when disturbed. A captive AMERICAN SHREW-MOLE climbed twigs and the side of its cage; it was also a powerful swimmer. This species constructs shallow runways and deep burrows. Surface activity takes place in a network of runways about four centimetres in diameter, just two centimetres below the top of the surface litter. The burrows are about two centimetres in diameter and run between 1 and 12 centimetres deep. Small ventilation ducts bring in air from the surface. Digging actions are very similar to those of other moles—alternate side-to-side movements of the front feet. Rather than pushing the dirt above ground to form molehills, the soil is pressed into the sides of the burrow. Having no well-defined activity period, AMERICAN SHREW-MOLES are active at all hours, taking intermittent brief rests lasting one to eight minutes. There are two reports of nests made of wood fragments situated above ground in rotten RED ALDER trees. The nests were reached by tunnels inside the tree.

Little information is available on actual population numbers, but densities of 12 to 15 per hectare were reported in ideal habitats in Washington. It seems to be a relatively uncommon species in BC; most researchers report less than one capture per 100 trap nights of effort. Dale Seip and colleagues, for example, captured only 48 AMERICAN SHREW-MOLES in 13,722 trap nights using pitfall traps. Similarly, of the 999 small mammals captured by Gustavo Zuleta and Carlos Galindo-Leal in the lower Fraser River valley, only 16 were AMERICAN SHREW-MOLES. Home range and movements have not been determined.

With a higher metabolic rate than would be expected for its body mass, AMERICAN SHREW-MOLE has large daily energy requirements and a voracious appetite. A 10-gram captive individual was observed to eat a 1.3-gram earthworm in 10 seconds; it consumed 1.4 times its body mass in 12 hours. The only information on food habits for BC are Kenneth Racey's observations of finding small insects, beetles and small worms in stomachs. In Oregon and Washington, earthworms are usually the major food accounting for 42 to 82 per cent of the stomach remains. Other prey eaten are insect larvae, adult beetles, grasshoppers, sowbugs (wood lice), snails, slugs and centipedes. Seeds and other plant material are eaten on occasion. A seasonal study in the Cascade Mountains of Washington found that, in September, invertebrates formed 75 to 88 per cent of the diet; in July, however, conifer seeds (36 per cent) and lichens (32 per cent) were the major foods and invertebrates (18 per cent) were minor. Captive AMERICAN SHREW-MOLES readily consumed conifer seeds, especially SITKA SPRUCE seeds, and various species of fungi.

Remarkably, AMERICAN SHREW-MOLE is completely blind and is dependent on its sense of touch to locate food and find its way. In addition to the sensitive bristles on the nose, it has bristle hairs on the tail that are probably sensitive to touch. The role of sound in the behaviour of this species is unknown. Walter Dalquest and Donald Orcutt found that it responded to sounds (8 to 30 kilohertz), suggesting that it may be adapted to hear some high-frequency sounds. More research is needed to determine its use of sound. The sense of smell is poorly developed, and the nose acts primarily as a tactile organ. When searching for food, it swings its nose from side to side, tapping the surface of the ground. Insect pupae and sowbugs are flipped over and pounced upon. Earthworms are bitten along their entire length and eaten whole or chewed into smaller pieces.

In Washington, AMERICAN SHREW-MOLE has a lengthy breeding season that begins in February and extends to late September, although few animals are in breeding condition after mid-May. The length of the gestation period has not been determined for this species. Females probably produce only one litter per year; the litter size ranges from one to four. Breeding data for BC are limited to information from a few museum specimens. Four pregnant females, with three or four embryos each, were taken between April 1 and June 15. Four nursing females were trapped between April 30 and June 12. These data are consistent with a breeding season that extends from March to June. Based on his capture of a female with four fully developed embryos on April 1, Kenneth Racey concluded that in BC the young are born in early April.

AMERICAN SHREW-MOLES are born naked with no vibrissae, no erupted teeth and no nails on their digits. Newborn young weigh less than a gram; their total length is about 26 millimetres and their tail length is about 5 millimetres. Growth and maturation of the young have not been studied, but they probably reach sexual maturity in the spring following their birth.

Because it is often active above ground, this species is more vulnerable to predators than COAST MOLE and TOWNSEND'S MOLE. Owls are probably the major predator (see photo on page 31). Garter snakes and a few mammals, such as NORTHERN RACCOON and DOMESTIC CATS, also prey on AMERICAN SHREW-MOLES.

CONSERVATION STATUS

AMERICAN SHREW-MOLE is secure in BC. It has a large range in southwestern BC and occupies many habitats. Local populations in some parts of the lower Fraser River valley may be at risk because of rapid habitat loss and fragmentation from urban development. For example, an isolated population occurs in Burns Bog in the heavily urbanized municipality of Delta.

SYSTEMATICS AND TAXONOMY

A molecular genetics study of the Talpidae based on mitochondrial and nuclear DNA revealed that the AMERICAN SHREW-MOLE is the only North American member of a genetic clade in the mole family that includes two species of Japanese shrew-moles and the LONG-TAILED MOLE of China.

Three to five subspecies of AMERICAN SHREW-MOLE have been recognized. There is disagreement on the number of subspecies in BC. Some taxonomists list two: *Neurotrichus gibbsii gibbsii* and *Neurotrichus gibbsii minor*. However, Terry Yates in his dissertation research assigned all BC populations to *Neurotrichus gibbsii gibbsii*, and we follow his taxonomy here. Subspecies of the AMERICAN SHREW-MOLE are based on size and pelage colour. Size variation is largely clinal with a south-to-north trend of decreasing size. No genetic studies have been done to assess the validity of these races.

- *Neurotrichus gibbsii gibbsii* (Baird)—Ranges from northern California throughout the Cascades of Oregon and Washington to British Columbia.

REMARKS

This is the only vertebrate known to have a pigmented layer covering the anterior surface of the eye lens. The AMERICAN SHREW-MOLE is functionally blind and shows no response to a bright light.

From the number of photos posted on iNaturalist, dead AMERICAN SHREW-MOLES are often encountered in BC. They are likely killed and then dropped by owls or mammalian predators. Their strong pungent odour may make them unpalatable to some predators.

The species name *gibbsii* is an eponym for George Gibbs, a naturalist with the Pacific Railroad Survey in the US, who collected the type specimen near Mount Rainier in 1854.

REFERENCES

Aubry, Crites and West (1991); Campbell and Hochachka (2000); Carraway and Verts (1991); Dalquest and Orcutt (1942); Gomez and Anthony (1998); Gunther, Horn and Babb (1983); He et al. (2017); Lewis (1983); Moon and Leonard (2001); Racey (1929); Seip and Savard (1991); Whitaker, Maser and Pedersen (1979); Yates (1978); Zuleta and Galindo-Leal (1994).

Coast Mole *Scapanus orarius*

DESCRIPTION

COAST MOLE has a chunky body with short limbs and a short, nearly naked tail. The pelage consists of short, soft, velvety fur that ranges from blackish-brown to grey dorsal fur and is paler with grey on the undersides. The winter pelage is darker, nearly black. The front feet are broad and shovel-like with long, flat claws; the hind feet are not enlarged and have short, weak claws. The snout is long and almost naked. The eyes are minute; external ears are absent. Adult males are larger than adult females.

Cranial/dental traits: skull with a broad braincase and long rostrum, complete zygomatic arches and complete auditory bullae; 44 teeth; incisors 3/3, canines 1/1, premolars 4/4, molars 3/3; first upper incisor long and broad; other upper incisors conical-shaped and similar in size; lower incisors, canine, and first, second and third premolar small and conical-shaped.

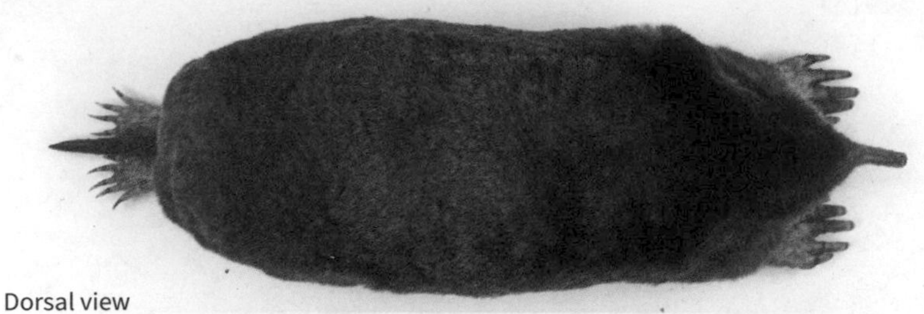
Dorsal view

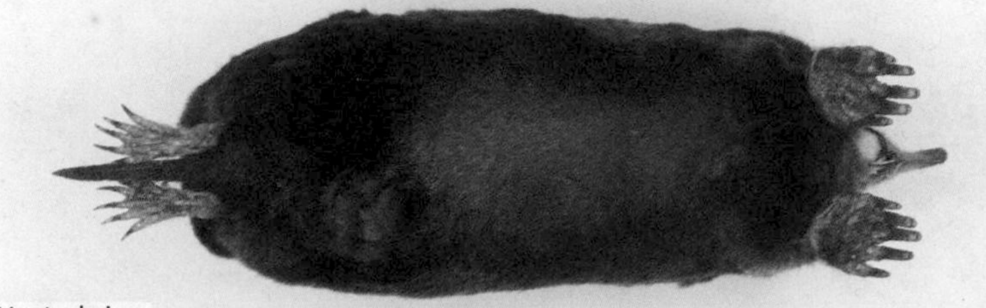
Ventral view

Mandible (dorsal view)

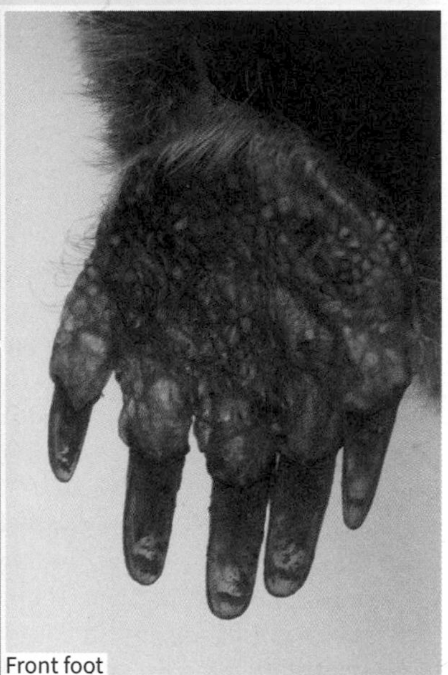
Front foot

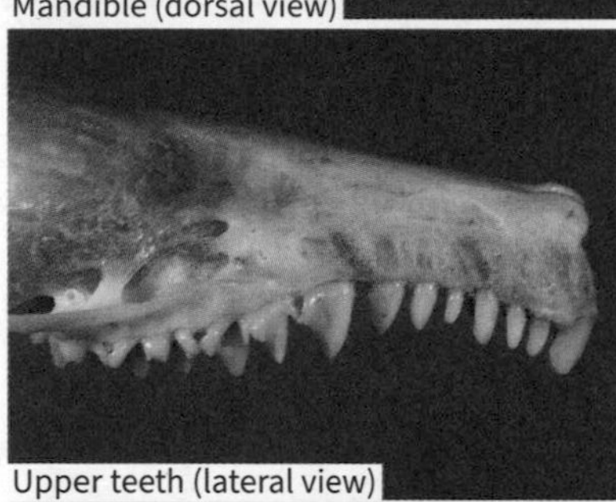
Upper teeth (lateral view)

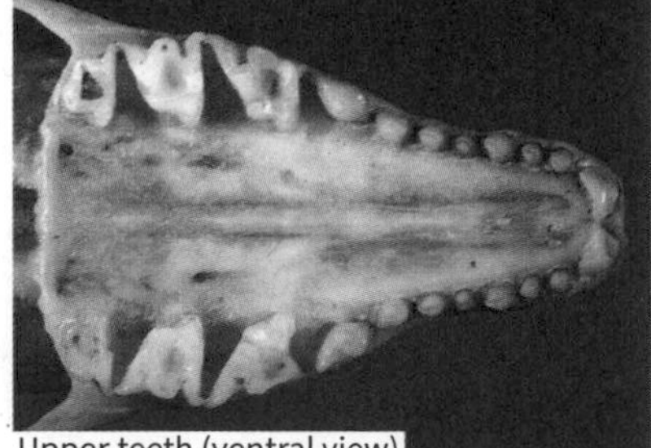
Upper teeth (ventral view)

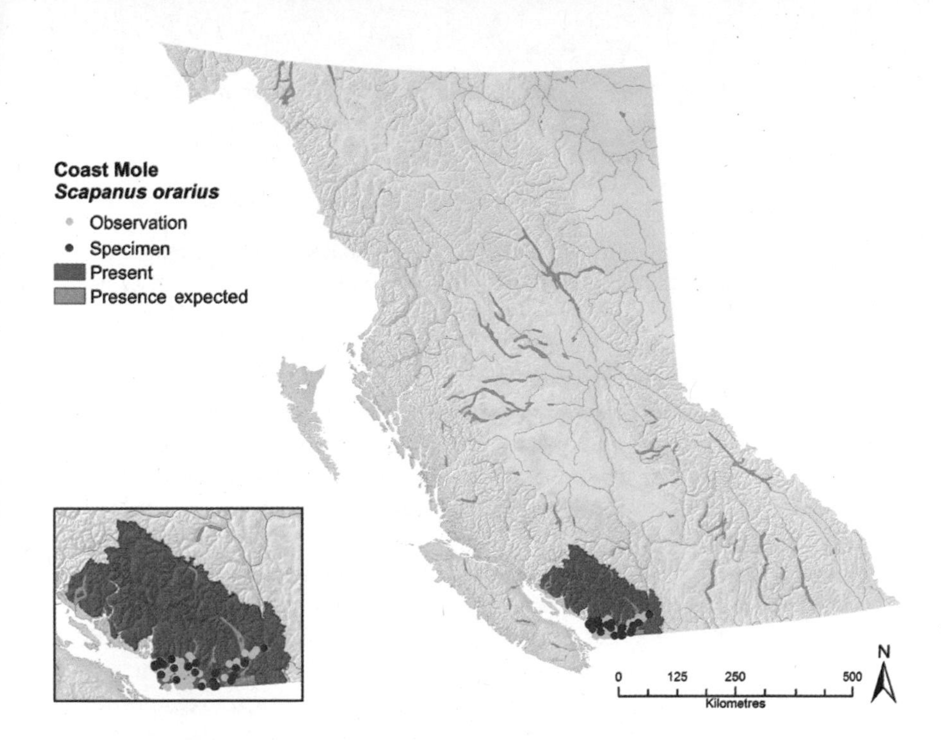

DISTRIBUTION

COAST MOLE inhabits western North America from northern California to BC; the eastern limits of its range are in extreme western Idaho. In BC, it is restricted to the lower Fraser River basin as far east as Hope and Agassiz and north to the south shore of Burrard Inlet.

MEASUREMENTS

	Mean	Range	Sample size
Total length:	162 mm	145–181 mm	*n* = 129
Tail vertebrae:	33 mm	28–41 mm	*n* = 121
Hind foot:	21 mm	18–24 mm	*n* = 128
Mass—females:	56.6 g	45.6–66.9 g	*n* = 15
Mass—males:	69.1 g	46.0–78.1 g	*n* = 16

MORPHOLOGICALLY SIMILAR SPECIES

The only species that could be misidentified as a COAST MOLE is TOWNSEND'S MOLE. Similar in external appearance and cranial traits, they can only be distinguished by size. TOWNSEND'S MOLE has a larger body (a total length greater than 175 millimetres, with the hind foot length greater than 24 millimetres) and skull (skull length greater than 37.0 millimetres). Measurements (see the TOWNSEND'S MOLE account) of the molehill and tunnel size may be used as an indirect method to identify the two species.

NATURAL HISTORY

The COAST MOLE exploits a much broader range of habitats than TOWNSEND'S MOLE, preferring moist but well-drained soils, including glacial tills, clays, river deposits, and the sands and gravels of benchland. It avoids flooded lands and acidic soils, such as peat bogs. Valentin Schaefer's research demonstrated that the greatest number of molehills occurred in loose soils with high water content and soils with large numbers of earthworms. Reduced mole activity in acidic soils probably reflects the scarcity of earthworms. Although most common in agricultural lands and riparian habitats along streams and rivers, the COAST MOLE is also found in forested areas. In BC it occurs at elevations below 350 metres.

Several types of tunnels are constructed. Hunting tunnels, appearing as surface runs or ridges on the soil, are made just below the surface of the ground. The mole simply pushes the dirt to the side so there are no mounds. Usually these tunnels are used only once for hunting, dispersal or searching for a mate during the breeding season. Most activity takes place in tunnels 7 to 20 centimetres below the surface. During dry spells, tunnels one to two metres deep may be constructed. A COAST MOLE tunnel two metres below the surface in compact sand was found during the excavation of a well at Agassiz. Tunnels are about four centimetres in diameter. Small enlarged chambers (10 by 8 centimetres)

Fresh mound made by a COAST MOLE.

occur at various intervals, often at the ends of tunnels or the junction of several tunnels. Presumably they serve as chambers for resting or nesting.

Soil from tunnel excavation is deposited in mounds or molehills on the surface. They are smaller than those made by TOWNSEND'S MOLE, averaging 30 centimetres in diameter and 11 centimetres high. Digging is most pronounced in autumn and winter when the soil is moist and easy for moles to work. This is the time of year when COAST MOLES expand their range into new areas. It has been estimated that between October and March, one individual can construct as many as 400 molehills. During winter, molehills are very conspicuous in fields, lawns, pastures, golf courses and grassy borders adjacent to highways.

During the breeding season, females construct nursery nests for rearing their young. Nests are situated at shallow depths (15 centimetres below the ground) where they are heated by the sun. They are about 20 centimetres across and are lined with dried grass. Nests typically have several entrances that are interconnected with tunnels. Unlike TOWNSEND'S MOLE, there are usually no large earth mounds associated with the nursery nests.

Several researchers have suggested that this mole is primarily nocturnal. However, an individual marked with a radioactive tracer and tracked with a Geiger counter was active throughout a 24-hour period with activity rhythms of about four hours. Subterranean movements of this animal ranged from 2 to 30 metres. The home range of the COAST MOLE averages about 0.12 hectares. Population estimates for the province are from less than one mole per hectare to as many as 13 per hectare in ideal habitats. Population numbers correlate with the physical characteristics of soil and the numbers of earthworms.

Observations of caged animals suggest that the COAST MOLE hunts almost continuously, using its snout to locate prey by scent or touch. It feeds predominantly on earthworms. The mole removes soil from the prey with its front feet, and then pulls the worm into its mouth, rapidly chewing and swallowing. It also eats snails, slugs, millipedes, centipedes and various soil insects; shrew and mouse remains have been found in a few stomachs. In an analysis of the stomach contents of 108 COAST MOLES trapped in BC, 93 per cent contained earthworms, 2 per cent slugs, 3 per cent insects and 2 per cent unknown material. Captive COAST MOLES avoided hairy caterpillars and dried earthworms. In captivity, it can eat nearly twice its body weight in earthworms daily. Wild moles may not consume as much, but they must take extremely large numbers of earthworms. It has been estimated that a population of 10 per hectare would eat some 219,000 earthworms per year.

Most literature suggests that COAST MOLE foraging is confined to underground tunnels, but there are several recent observations in California of adults foraging above ground at night. One COAST MOLE was observed above ground eating a moth caterpillar. Above-ground foraging movements are usually at night located short distances from burrows to minimize any risk of predation.

In BC, mating takes place from January to early March. Testes begin to enlarge in early January, and by May they have regressed in size. Females with embryos have been found from February to April; by May all females have given birth. The gestation period is unknown. Females produce one litter a year and are capable of breeding in their first year. The average litter size is four, although females in their first year usually have only two young. Newborn young have not been observed, but presumably they are naked and lack teeth. Two young COAST MOLES (13.0 and 15.0 grams) found in a nest were almost naked but had well-formed feet and tails. Estimates based on tooth wear as an indicator of age suggest that COAST MOLES may live three to four years in the wild.

Major predators are owls, especially BARN OWLS, which generally prey on juvenile moles when they leave the breeding nest and disperse above ground in summer. In Oregon, the RUBBER BOA is a predator of young moles in their natal nest. DOMESTIC DOGS and DOMESTIC CATS are also major predators, especially during the period of juvenile dispersal. Winter and spring flooding may have a detrimental effect on some populations. During a major flood of the Fraser River in 1948, COAST MOLES were observed escaping from fields and swimming to high ground. Mole populations declined in the flooded areas, but this may have resulted more from the destruction of earthworms than the direct loss of moles from drowning.

CONSERVATION STATUS

The COAST MOLE is secure in BC and considered by many to be a pest species. It occupies a part of the province that is undergoing rapid development and habitat loss; however, populations seem to persist in golf courses, agricultural lands and lawns or gardens of residential areas. The impact of trapping and poisoning on local populations has not been studied.

SYSTEMATICS AND TAXONOMY

A molecular genetics study of the Talpidae that was based on mitochondrial and nuclear DNA revealed that COAST MOLE is a member of a North American clade in the mole family comprising the EASTERN MOLE and the four species of *Scapanus* (which includes TOWNSEND'S MOLE, the closest relative to the COAST MOLE). Although morphologically similar differing mostly in size, DNA data suggest that they separated in the late Pliocene epoch several million years ago.

Two subspecies are recognized; one is found in BC. No genetic studies have been done assessing genetic variation among populations of this species.

- *Scapanus orarius schefferi* Jackson. A large form occupying eastern Oregon, Washington, western Idaho and BC.

REMARKS

This species is sufficiently abundant around gardens, farms and golf courses to be regarded as an agricultural pest in the province; several commercial pest-control companies specialize in trapping moles. Although their tunnels may damage the roots of plants, much of the damage is esthetic and the pest status of this mole may be exaggerated.

There are no moles native to any BC island. In November 2011, a COAST MOLE was trapped on the library grounds of the University of Victoria campus. It was a stowaway in a root ball of nursery stock that came from the Lower Mainland of BC. The mole now resides in the research collections of the Royal BC Museum.

The species name *orarius* is from Latin, meaning "coasting" or "along the coast."

REFERENCES

Campbell, Manuwal and Harestad (1987); Dinets (2017); Glendenning (1959); Hartman and Yates (1985); He et al. (2017); Schaefer (1978); Schaefer and Sadleir (1981); Sheehan and Galindo-Leal (1996, 1997); Whitaker, Maser and Pedersen (1979); Yates (1978).

Townsend's Mole *Scapanus townsendii*

DESCRIPTION

TOWNSEND'S MOLE is a large version of the COAST MOLE in appearance. Its fur is short, soft and velvety, ranging in colour from blackish brown to grey, and darker in winter pelage. In some populations, a few individuals have irregular white or yellow markings on their undersides. We did not observe these markings on the few specimens available from BC. The head has minute eyes and no external ears; the snout is long and almost naked. TOWNSEND'S MOLE has broad and shovel-like front feet with long flat claws. The hind feet are not enlarged and have short weak claws, and the tail is short and nearly naked. Adult males are larger than adult females.

Cranial/dental traits: skull with a broad braincase and long rostrum, complete zygomatic arches and complete auditory bullae; 44 teeth; incisors 3/3, canines 1/1, premolars 4/4, molars 3/3; first upper incisor long and broad; other upper incisors conical-shaped and similar in size; lower incisors, canine, and first, second and third premolar small and conical-shaped.

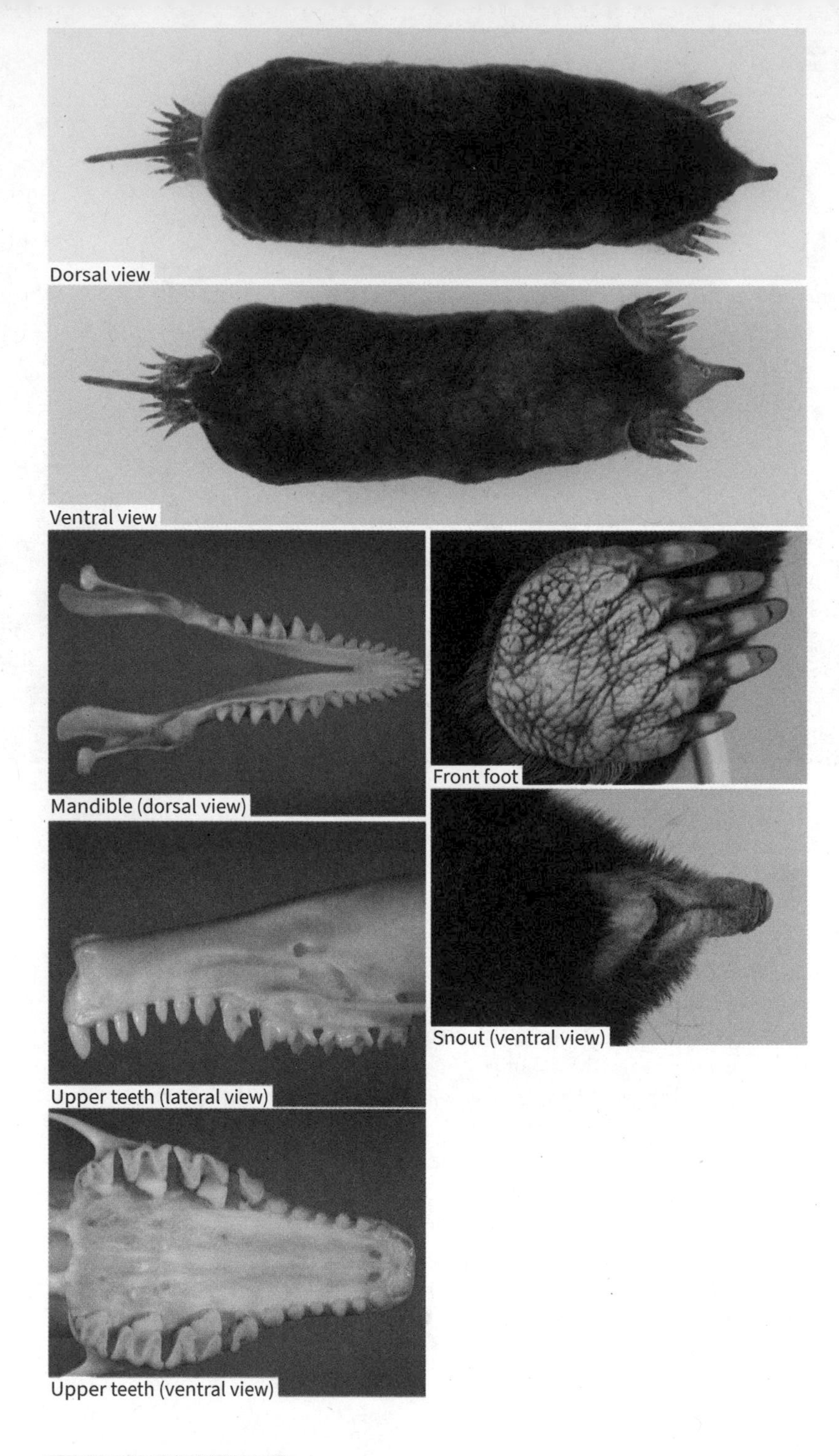
Dorsal view
Ventral view
Mandible (dorsal view)
Front foot
Upper teeth (lateral view)
Snout (ventral view)
Upper teeth (ventral view)

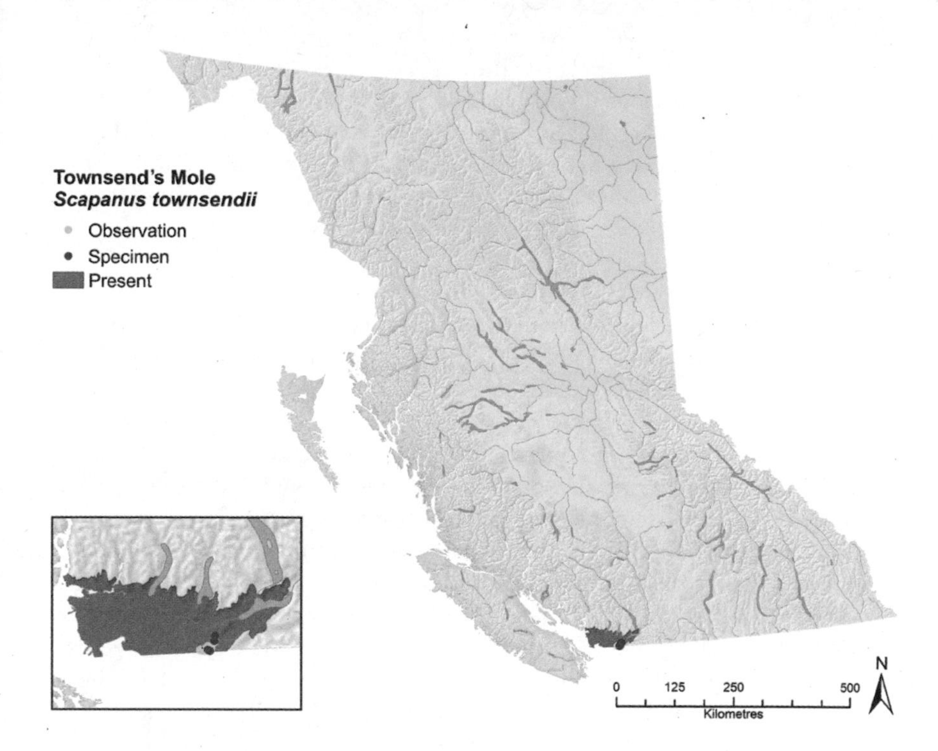

DISTRIBUTION

TOWNSEND'S MOLE inhabits the Pacific coastal regions of northern California, Oregon and Washington. In BC, where it reaches the northern edge of its range, it has one of the smallest distributional areas of any mammal in the province confined to a localized area around Huntingdon, adjacent to the international border, and a small area east of Abbotsford north of Highway 1. The range extent is only 20 square kilometres, and the area actually occupied within this range extent is only 13 square kilometres.

MEASUREMENTS

	Mean	Range	Sample size
Total length:	205 mm	179–237 mm	$n = 30$
Tail vertebrae:	38 mm	31–45 mm	$n = 30$
Hind foot:	25 mm	23–29 mm	$n = 31$
Mass—females:	113.5 g	96.0–122.0 g	$n = 5$
Mass—males:	137.9 g	121.0–164.0 g	$n = 7$

MORPHOLOGICALLY SIMILAR SPECIES

Only the COAST MOLE could be confused with TOWNSEND'S MOLE. See the COAST MOLE species account for diagnostic external and cranial traits. Tim Sheehan and Carlos Galindo-Leal found that the size of molehills made by the two species differ, with those made by TOWNSEND'S MOLE usually more than 15 centimetres high and 40 centimetres wide.

NATURAL HISTORY

TOWNSEND'S MOLE typically inhabits lowland meadows, cultivated fields, floodplains and prairie habitats. High-elevation populations are found in the Olympic and Cascade Mountains of Washington and Oregon, but the BC population occupies low-elevation agricultural land in the Fraser River valley. Most occurrences in BC are in hayfields, pastures, lawns and gardens that are associated with deep, dry silty loam soils (Marble Hill and Ryder soil types) that are not prone to flooding. They are ideal soils for digging and maintaining subterranean tunnels and nests and yield abundant earthworms, the major food source of TOWNSEND'S MOLE. They are soil types with a very limited distribution in Fraser River valley, and this may account for this species's restricted range.

A proficient digger, TOWNSEND'S MOLE constructs several types of tunnels. It makes shallow tunnels (5 to 15 centimetres below ground) for foraging and possibly for locating mates in the breeding season. Usually the surface tunnels are used only once. Molehills are deposited on the surface along these tunnels. TOWNSEND'S MOLE also digs permanent deep tunnels, usually 10 to 20 centimetres below the surface, although it may construct deeper tunnels (one to three metres deep) under roads, buildings and uncultivated areas along fences. Excavated soil is deposited as conical mounds above these tunnel systems. The average molehill is about 43 centimetres in diameter and 17 centimetres high.

Potential habitat near Huntingdon, BC.

In the breeding season, females construct nursery nests in underground cavities where they give birth to and raise their young. The nest chambers are situated 15 to 20 centimetres below ground and are about 23 centimetres long and 15 centimetres high. Dirt from excavating these chambers is usually deposited on the surface in a large mound 70 to 130 centimetres in diameter and 30 to 45 centimetres high. Each nest chamber has 3 to 11 lateral tunnels and an escape tunnel that runs from the bottom of the nest to another tunnel or burrow. The nest has an outer layer constructed from coarse grass, moss or leaves, and an inner layer made from fine dry grass. Green plant material is often used for the outer layer. Some nursery nests are used only one year, and others may be reused for several breeding seasons. Until a female gives birth, she will readily abandon her nest if disturbed and build a new nest within four or five days.

Recaptures of marked TOWNSEND'S MOLES suggest that subterranean movements of adults with established burrow systems are limited: distances between capture sites ranged from 3 to 116 metres, and the average distance between captures was about 40 metres. Long-distance movements (beyond 100 metres) are usually undertaken in the dry months of summer in poor habitats where earthworms are scarce. Recent observations in California suggest that this mole does some surface foraging on rainy or foggy nights in winter. When dispersing from the nursery nest in late spring and summer, young TOWNSEND'S MOLES move considerable distances (up to 800 metres) above ground.

This species possesses a strong homing ability. Individuals displaced short distances (100 to 200 metres) by annual floods usually reoccupy their original tunnel systems quickly after flooding. Moles displaced artificially have successfully returned to their original locations from distances of up to 450 metres. Although TOWNSEND'S MOLES are strong swimmers, canals and rivers can be major barriers to their movement.

In Oregon, population densities reach 12 per hectare in ideal habitats; as many as 805 mounds per hectare have been counted. Densities may be as low as 0.4 per hectare in areas with few earthworms or unsuitable soil. The only information for BC is population estimates of 3.2 and 5.2 per hectare for two fields west of Huntingdon studied by Tim Sheehan.

The diet is predominantly earthworms. In Oregon several studies based on stomach contents revealed that 70 to 90 per cent of the prey is earthworms. Small amounts of other invertebrates (centipedes, millipedes, snails, slugs and insects) and a few small mammals (shrews or mice) have also been identified in stomach remains. Vegetation (bulbs, roots, grass, carrots, parsnips, oats and beans) is also eaten. Probably because of its more restricted habitat, the diet appears less diverse than that of the COAST MOLE.

In Oregon the breeding season begins in early winter; males with enlarged testes have been found in November. By February the testes have begun to decrease in size, and by mid-March males are no longer in breeding condition.

Pregnant females have been observed as early as mid-March and no later than mid-April. Female TOWNSEND'S MOLES appear to have only one litter per year. Reproductive data for the BC population is limited to anecdotal information for a few museum specimens: a male with enlarged testes was taken on March 24, and a female with a fetus about 20 per cent developed was captured on April 27. The meagre data suggest a later breeding season in BC than in Oregon. The gestation period has not been determined for this species, but it is assumed to be four to six weeks.

Females bear one to four young, with three most common. Newborn moles are naked and pink and lack teeth and distinguishable eyes; they weigh only 5.0 grams. By 10 days, their skin colour has changed to grey; the fur begins to grow within three weeks of birth. They are completely furred by 30 days, and weigh 60.0 to 80.0 grams. Within a month of weaning, young begin to leave their mother's nest and disperse. In Oregon, they disperse in May and June. TOWNSEND'S MOLES are capable of breeding in their first winter after birth.

DOMESTIC DOGS and DOMESTIC CATS are opportunistic predators of TOWNSEND'S MOLES, but the major predators are owls, particularly BARN OWLS. Most susceptible to predation are the young-of-the-year when dispersing in summer above ground at night. They will cross roads when dispersing, and a few have been found dead on highways in summer. In the dairy-farming regions of Oregon, young moles are killed by cattle trampling the nursery nests. In lowland areas, winter flooding also takes its toll. Richard Giger counted 62 TOWNSEND'S MOLES that had died in a severe January flood in his study area in Oregon. Moles may be trapped in their tunnels or die from exhaustion when attempting to swim to higher ground.

CONSERVATION STATUS

COSEWIC assessed this species as endangered in Canada, and it is listed under the Species at Risk Act. It also is on the BC Red List. A small distributional area and presumably small population limited by a specific soil type, loss or degradation of habitat from development or agricultural activities, and direct killing as a pest are factors contributing to its designation. All of the known occurrences in BC are on private land; none are in protected areas. A national recovery plan released in 2015 delimited 18 square kilometres within the known range extent as critical habitat. The largest portion of critical habitat was located west of Huntingdon and south of Highway 1. With development and the conversion of pasture land to berry-crop production, it seems likely that some critical habitat mapped in the 2015 recovery strategy has been lost. A visit to the Huntingdon area in 2022 by one of us (Nagorsen) to photograph habitat for this book revealed that extensive areas of former pasture land have been converted to commercial blueberry production. Fortunately, the

November 2021 flood of the nearby Sumas Prairie area was peripheral to this mole's range extent.

SYSTEMATICS AND TAXONOMY

A molecular genetics study of the Talpidae based on mitochondrial and nuclear DNA revealed that TOWNSEND'S MOLE is a member of a North American clade comprising the EASTERN MOLE and the four species of *Scapanus* including the COAST MOLE, its closest relative. Two subspecies are recognized with one found in BC. No genetic studies have been done assessing genetic variation among populations or the validity of the two subspecies.

- *Scapanus townsendii townsendii* (Bachman). A widespread race occupying the entire range of the species except for high elevations in the Olympic Mountains of Washington where it is replaced by another subspecies *Scapanus townsendii olympicus*.

REMARKS

Information on the BC population was limited to historical data associated with museum specimens collected 40 to 90 years ago at the old Racey farm near Huntingdon until Tim Sheehan initiated a field study in 1994. He confirmed that a breeding population still existed near Huntingdon and found an additional local population east of Abbotsford north of Highway 1. As background for the 2003 COSEWIC status assessment, Valentin Schaefer re-surveyed the known range extent using molehill size to identify TOWNSEND'S MOLE territories. No overall monitoring or inventory has been done since his 2003 work. A systematic inventory of known territories and the condition of their habitat is essential to determine the current population status.

The survey for the 2003 COSEWIC report found a number of COAST MOLE territories within the range extent of the TOWNSEND'S MOLE. Competition between these two mole species in BC has not been studied.

The species name *townsendii* is an eponym for James Kirk Townsend, whose name is associated with many bird and mammal species that he collected in the lower Columbia River region in the 1830s.

REFERENCES

BC Ministry of Environment (2014); Carraway, Alexander and Verts (1993); COSEWIC (2003); Dinets (2017); Environment Canada (2016); Giger (1973); He et al. (2017); Kuhn, Wick and Pedersen (1966); Pedersen (1963, 1966); Sheehan and Galindo-Leal (1996, 1997); Whitaker, Maser and Pedersen (1979).

Arctic Shrew

Sorex arcticus
Other common names:
Black-backed Shrew, Saddle-back Shrew

DESCRIPTION

Our most attractive shrew, ARCTIC SHREW has a distinctive tricoloured saddlebacked pelage characterized by a dark back that contrasts strikingly with the paler sides and undersides. The back ranges from dark brown to black, the sides are a paler brown and the undersides are grey to light brown. The tail is indistinctly bicoloured, brown to black above and paler on the underside. In winter, the pelage is longer, the sides and underside lighter than summer and the tail more distinctly bicoloured. Young-of-the-year are dull and bicoloured. They acquire the distinctive tricoloured pattern at their autumn moult, retaining this pelage throughout their lives. Captures in the Peace River area taken from mid-May to mid-June were still in their winter pelage.

Cranial/dental traits: 32 teeth; incisors 1/1, unicuspids 5/1, premolars 1/1, molars 3/3; upper incisor with a small- to medium-sized medial tine positioned within the pigmented area; third upper unicuspid taller than the fourth; a lingual pigmented ridge on upper unicuspids does not extend to the cingulum and is separated from the cingulum by a longitudinal groove; lower half of the ridge unpigmented; edge of pigment on anterior face of upper incisor sits at or below the pigment on its lateral cusp; hypocone on first upper molar usually pigmented; lower incisor has three denticles with pigment on the labial side continuous to the first, second or third denticle, with pigment on lingual side only extending to first denticle; both dentaries with small to large postmandibular foramen.

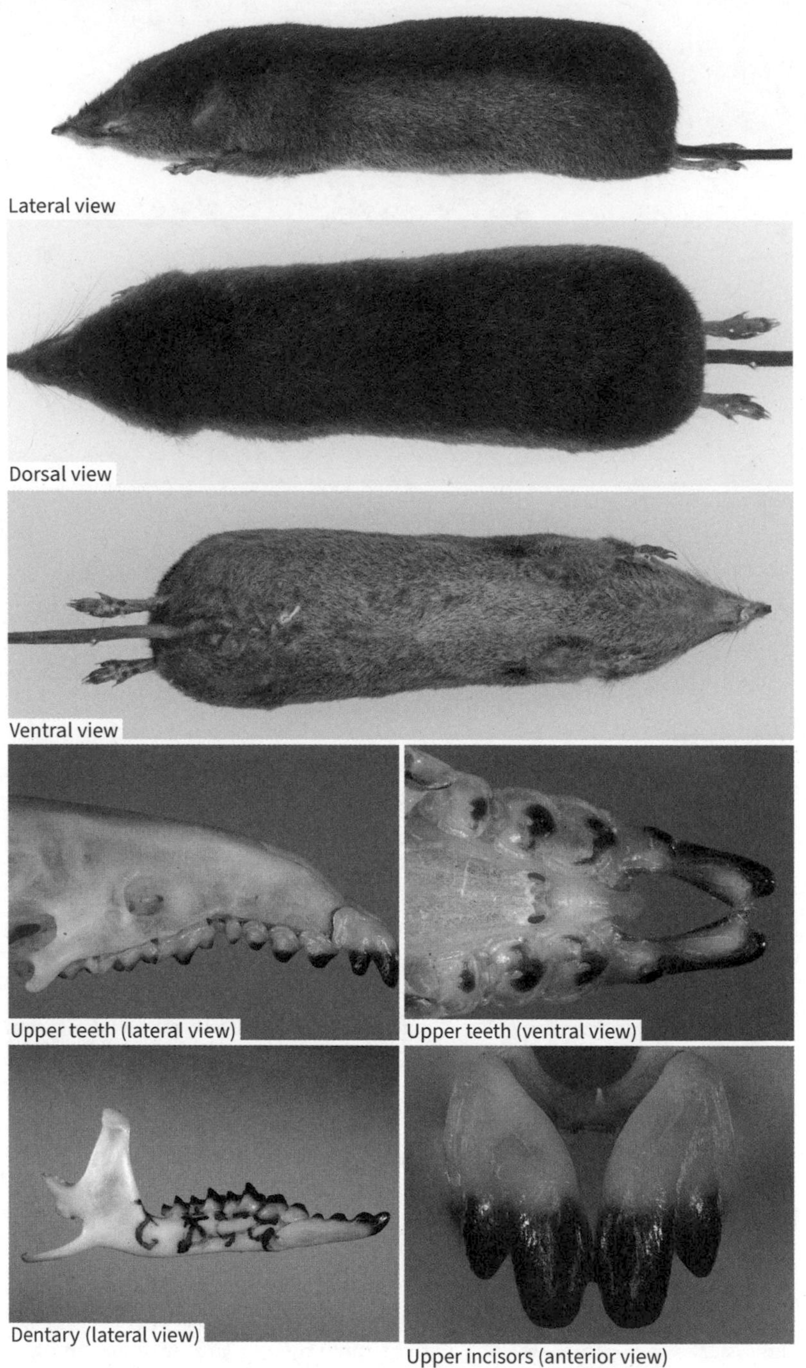
Lateral view
Dorsal view
Ventral view
Upper teeth (lateral view)
Upper teeth (ventral view)
Dentary (lateral view)
Upper incisors (anterior view)

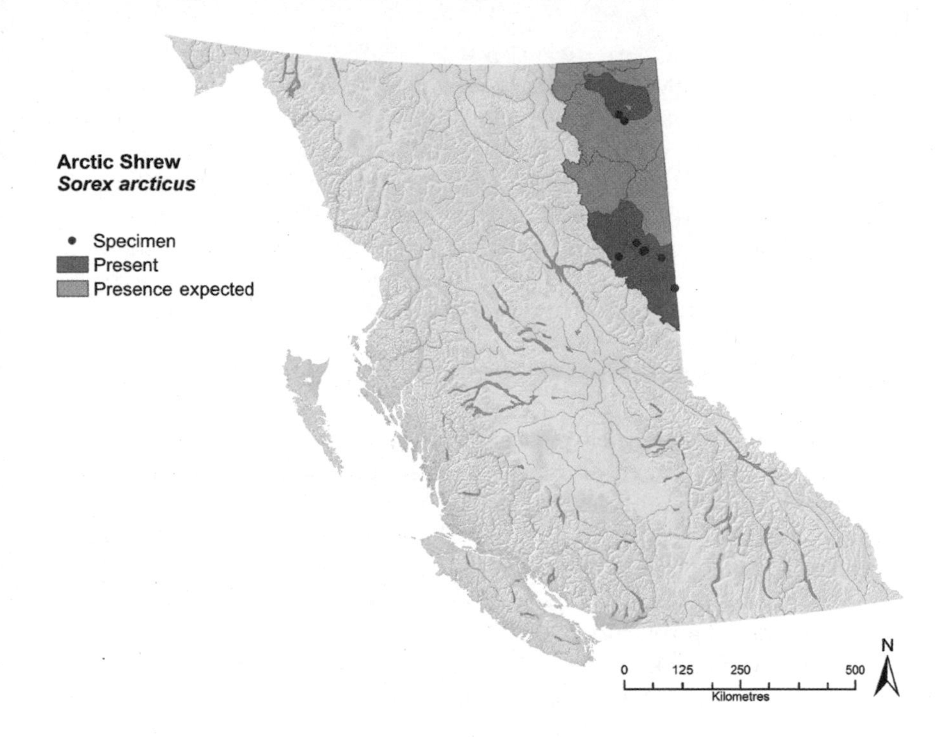

DISTRIBUTION

ARCTIC SHREW ranges across the boreal forests of Canada and the North Central United States. In BC, where it reaches the western limits of its range, it occupies the northeastern part of the province east of the Rocky Mountains from the Peace River north to Kotcho Lake. Although known from only eight locality records in the province, it is probably widespread in the Boreal Plains and Taiga Plains ecoprovinces.

MEASUREMENTS

	Mean	Range	Sample size
Total length:	116 mm	100–125 mm	$n = 30$
Tail vertebrae:	41 mm	36–46 mm	$n = 32$
Hind foot:	14 mm	11–19 mm	$n = 31$
Mass:	9.4 g	6.0–13.0 g	$n = 10$

MORPHOLOGICALLY SIMILAR SPECIES

ARCTIC SHREW closely resembles TUNDRA SHREW, but they are widely separated in their BC range. Species that co-occur with ARCTIC SHREW likely to cause misidentification are CINEREUS SHREW and DUSKY SHREW, two shrews that lack a strong tricoloured pelage. CINEREUS SHREW shares the pattern of a third upper unicuspid larger than the fourth, but it is smaller, usually weighing less than 6.0 grams and with a skull length less than 16.0 millimetres. The upper unicuspids on individuals with unworn teeth show a distinct pigmented ridge extending to the cingulum. DUSKY SHREW has a third upper incisor that is smaller than the fourth and upper unicuspids with a pigmented ridge extending to the cingulum.

NATURAL HISTORY

In eastern North America, ARCTIC SHREWS are most common in wet grass-sedge marshes or shrub thickets of willow and alder near lakes, beaver ponds and streams. It is also found in BLACK SPRUCE–TAMARACK bogs. Localized in its distribution, it is often confined to small pockets of suitable habitat. Although most habitat descriptions suggest a preference for moist non-forest habitats, a study in Alberta found that the majority of captures were in dry meadows and well-drained mixed forests.

In BC, ARCTIC SHREWS inhabit lowland forests and plateaus (below 760 metres). The only habitat descriptions available for the province are from Ian McTaggart Cowan's 1938 survey of the vertebrate fauna of the Peace River area. At Tupper Creek, he found ARCTIC SHREWS in MEADOW VOLE runways in sedge habitats that bordered marshland; at Charlie Lake it was captured in wet mossy habitats in mature TREMBLING ASPEN forest. McTaggart Cowan reported that ARCTIC SHREW was the most abundant shrew in his Peace River district study, but he did no sampling in the actual valley of the Peace River. In an unpublished 2011 study for BC Hydro, Les Gyug and Nagorsen failed to capture a single ARCTIC SHREW in their surveys for the Site C hydro dam at 24 sites located between the Moberly River and Hudson's Hope in the Peace River valley.

The only estimates of population numbers and movements are based on a six-year study of a population living in a TAMARACK bog in Manitoba. Population estimates ranged from about 10 animals per hectare in a peak year to less than 1 per hectare in a low year. Populations demonstrate the typical pattern for shrews, with low numbers in spring, increasing to a peak in August and September and declining in the fall. The average home range in TAMARACK bogs was about 0.6 hectares. Home range was not related to age or sex, but when populations were high, it decreased and there was considerable overlap in the daily movements of individuals.

In northeastern BC, this species coexists with CINEREUS SHREW, DUSKY SHREW and WESTERN PYGMY SHREW. Interactions among these species have not been not studied, but there is some evidence from eastern Canada that the lowest densities of ARCTIC SHREWS coincide with years when populations of CINEREUS SHREWS are high.

The ARCTIC SHREW is active at any time, but it tends to be more active in darkness. Active periods are short, averaging about three minutes. When resting, ARCTIC SHREWS sleep on their sides or rolled up with the head tucked under the body.

In captivity, ARCTIC SHREWS eat earthworms, spiders, grasshoppers, the larvae of moths and butterflies, flies, LARCH SAWFLIES and adult beetles, but appear to avoid ants. They also readily consume carrion, such as dead voles. The stomachs of wild animals that were trapped in TAMARACK bogs in Manitoba contained mostly insect remains. It is also a major predator of LARCH SAWFLY cocoons; in autumn this insect accounted for 70 per cent of the diet of the Manitoba population. A single ARCTIC SHREW may consume as many as 120 sawfly pupae in a day.

ARCTIC SHREWS have been seen climbing about 30 centimetres up shrubs and grass when hunting grasshoppers; they would pounce upon the grasshoppers and grasp them in their jaws and feet. A captive animal killed three grasshoppers in several minutes. Two wild ARCTIC SHREWS killed 33 grasshoppers in about 15 minutes.

The breeding season extends from late winter through summer. Pregnant females have been reported as early as April in some regions, but in most populations the first young appear in June. Embryo counts range from 5 to 11 and average about 8. Females may produce more than one litter per year. Breeding data for BC are scanty but are consistent with the pattern in eastern North America. A nursing female was captured at Charlie Lake on May 16, and two pregnant females with seven and eight embryos were taken at Kotcho Lake in mid-June. Females are capable of breeding by four or five months of age; however, the extent that ARCTIC SHREWS breed in their first summer varies among populations. In the southern parts of the range, breeding by young-of-the-year is rare; in Manitoba, however, young from the first litter typically breed in the summer of their birth. Mortality is highest in young animals, especially in the first month; it has been estimated that only 20 per cent of the young will survive beyond four or five months. The maximum life span for this species is about 18 months.

CONSERVATION STATUS

Although known from only a few occurrences, ARCTIC SHREW is considered secure by the BC Conservation Data Centre. Nonetheless, its range is in a region undergoing significant resource development and habitat change from oil and gas exploration, flooding from the Site C hydroelectric dam, forest harvesting and construction of wind-energy sites.

SYSTEMATICS AND TAXONOMY

ARCTIC SHREW is a member of the *Sorex araneus* group, comprising about a dozen species most of which inhabit Eurasia. The closest relative is the MARITIME SHREW of eastern Canada, which was originally treated as a subspecies of ARCTIC SHREW. Results from mitochondrial DNA suggest that the ancestor of the ARCTIC SHREW–MARITIME SHREW lineage crossed into North America from Asia by a land bridge several million years ago. Two subspecies are currently recognized for ARCTIC SHREW; one occurs in BC. They show few differences in morphology, and mitochondrial DNA sequences from a small sample suggest minor differences. A comprehensive genetic study is required to assess the validity of the two races.

- *Sorex arcticus arcticus* Kerr. A widespread race that ranges across western Canada to eastern Quebec.

REMARKS

Most of the biological information for ARCTIC SHREW is derived from research done in central and eastern North America. What little we know about its biology in BC is based on Ian McTaggart Cowan's observations in the Peace River region made 84 years ago! It is a good example of a BC shrew species in need of a focused research study.

The species name *arcticus* refers to "northern" or "arctic." Although certainly a northern shrew, almost all of its range is south of the Arctic Circle in the northern boreal forest.

REFERENCES

Baird, Timm and Nordquist (1983); Buckner (1957, 1964, 1966, 1970); Clough (1963); Kirkland and Schmidt (1996); Mackiewicz et al. (2017); McTaggart Cowan (1939); Salt (2005); Stewart, Perry and Fumagilli (2002); Van Zyll de Jong (1983b); Wrigley, Dubois and Copland (1979).

Pacific Water Shrew

Sorex bendirii

Other common names:
Marsh Shrew, Bendire's Shrew

DESCRIPTION

One of the largest shrews in the province, PACIFIC WATER SHREW is an attractive mammal with dorsal fur that varies from dark brown to black, dark-brown sides and ventral fur slightly paler than the dorsal fur. In winter pelage, the ventral fur is darker, nearly the same colour as the greyish dorsal pelage. The tail is a unicoloured dark brown above and below. The moult to summer pelage likely begins in April or May, with most individuals in complete summer pelage by July. The moult to the winter pelage begins in early September. The sides of the feet have a stiff fringe of hairs about one millimetre long, although the hairs may be missing on old adults with worn pelage.

Cranial/dental traits: skull large with rostrum down-curved in side profile; 32 teeth; incisors 1/1, unicuspids 5/1, premolars 1/1, molars 3/3; upper incisor with a small to large medial tine positioned at or within the pigmented region on the face of the incisor; third upper unicuspid shorter or equal in size to the fourth; upper unicuspids with a weak to strongly pigmented ridge that extends to the cingulum; lower incisor has three denticles with pigment area on the labial side covering all denticles, with continuous pigment on lingual side extending to the second or third denticle; postmandibular foramen rare, and if present, small and usually occurs in only one dentary.

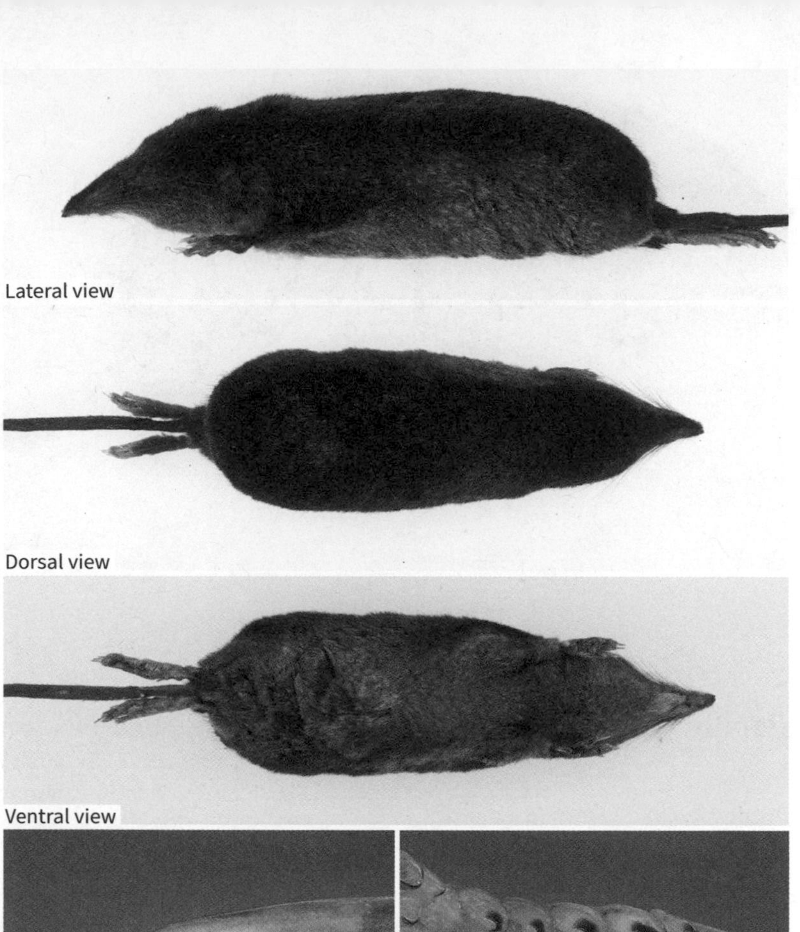

Lateral view

Dorsal view

Ventral view

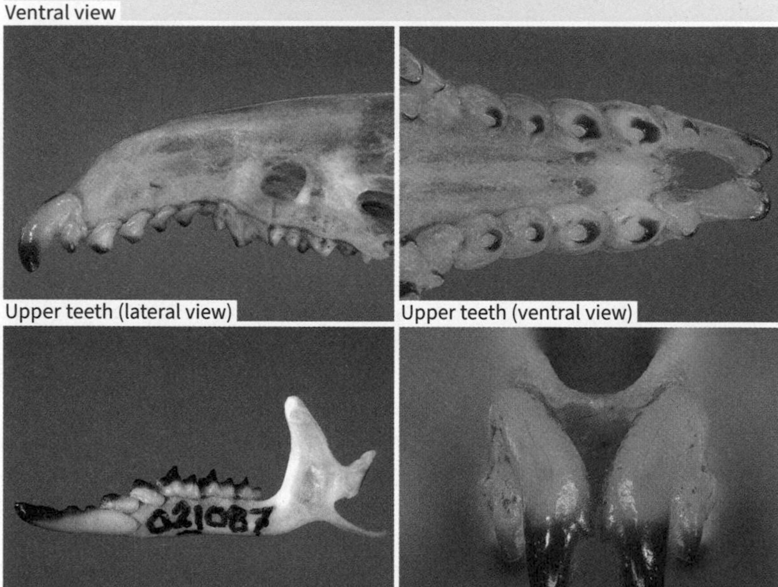

Upper teeth (lateral view)

Upper teeth (ventral view)

Dentary (lateral view)

Upper incisors (anterior view)

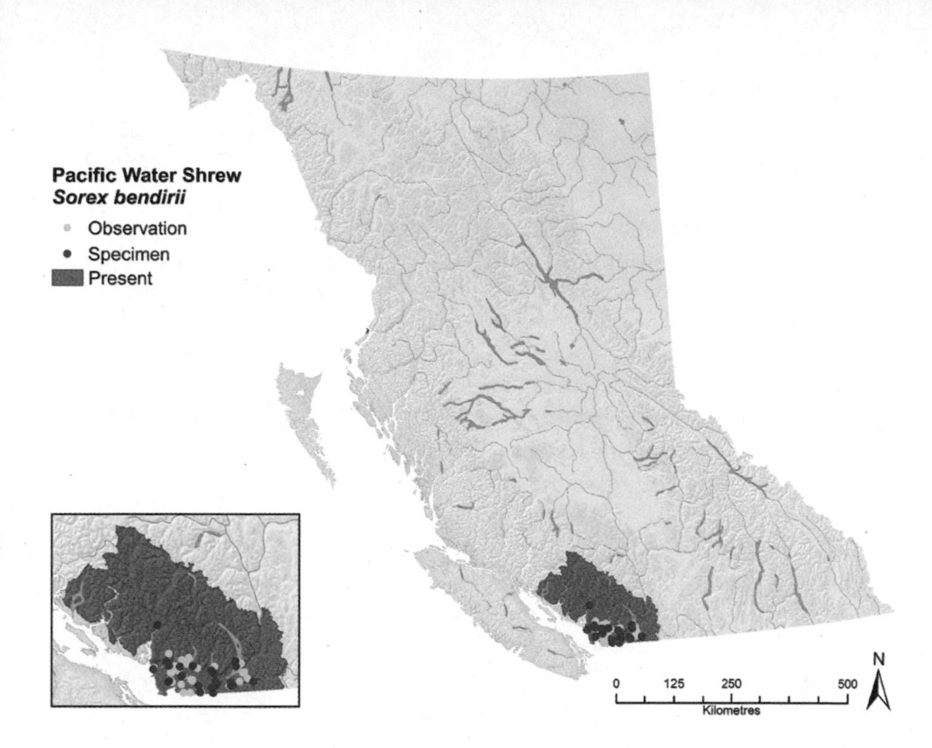

DISTRIBUTION

PACIFIC WATER SHREW inhabits the coastal lowlands and mountains of northern California, Oregon, Washington and BC, where it is restricted to the lower Fraser River basin. Eastern limits of its range are the Chilliwack valley and Agassiz; northernmost records are low elevations in the Coast Mountains on the north shore of Burrard Inlet, although a capture at Thunderbird Creek in Squamish in 2008 extends the northern limits of its known range. Eastern boundary of the BC distribution are unknown, but we suspect it may reach the Skagit River valley.

MEASUREMENTS

	Mean	Range	Sample size
Total length:	153 mm	130–176 mm	$n = 108$
Tail vertebrae:	69 mm	58–81 mm	$n = 108$
Hind foot:	19 mm	16–21 mm	$n = 108$
Mass—females:	11.7 g	10.4–13.2 g	$n = 5$
Mass—males:	13.6 g	10.0–17.2 g	$n = 14$

MORPHOLOGICALLY SIMILAR SPECIES

A large body size, dark-brown or black dorsal and dark-brown ventral fur, a fringe of stiff hairs on the hind feet and a skull length greater than 19.0 millimetres are the features that discriminate the PACIFIC WATER SHREW from all other shrews found within its range except for WESTERN WATER SHREW. External traits that identify WESTERN WATER SHREW are dark-grey to black dorsal fur with a silver-grey belly, a bicoloured tail that is paler ventrally, the ventral surface of its tail for about the last centimetre from the tip having a raised fleshy ridge or keel about one millimetre high fringed by stiff hairs, and longer, more conspicuous fringe of hairs on its hind feet. Although smaller in size than PACIFIC WATER SHREW, the body and skull measurements of the two species overlap to some extent. The skull of the WESTERN WATER SHREW lacks the down-curved rostrum.

NATURAL HISTORY

A habitat specialist, PACIFIC WATER SHREW occurs in moist riparian habitats associated with stream sides and marshes. Its elevational range in BC is unclear. Most occurrences are at low elevations, but it has been captured at 680 metres elevation in the Chilliwack valley and a historical museum specimen was taken at 850 metres elevation on Mount Seymour (no specific location) in the North Shore mountains. A recent study in Washington State by Jordan Ryckman found PACIFIC WATER SHREWS at 700 to 850 metres elevation on the eastern side of the Cascades near Snoqualmie Pass.

Typically this shrew is associated with forests of RED ALDER, BIGLEAF MAPLE, WESTERN HEMLOCK or WESTERN REDCEDAR that border streams and SKUNK CABBAGE marshes. These habitats usually have an extensive canopy cover, abundant shrubs and a ground cover of woody debris and fallen logs. Critical biophysical features listed in a recovery strategy by Environment Canada were coniferous or deciduous forest or dense wetland vegetation that provides cover and maintains a moist microenvironment, an area of water to support foraging and the moist microenvironment, cover from downed wood, and material for nesting and foraging. PACIFIC WATER SHREW is less dependent on standing water than WESTERN WATER SHREW, the only shrew found in the lower Fraser valley with similar aquatic habits. PACIFIC WATER SHREW has been captured in dry forests well away from water, but during the dry summer months, it is usually found within 200 metres of streams. Populations may occur in localized patches of suitable microhabitat. At Point Grey, near the University of British Columbia, nine PACIFIC WATER SHREWS were captured by Mary Jackson just above the tide line in beach debris near small pools created by springs.

PACIFIC WATER SHREW's relationship with forest age is not clear. Several studies in Oregon demonstrated that this species was most abundant in old forests; but a study in the Cascade Mountains of Washington found it most

Capture site near Chilliwack.

common in younger forests. Regardless of forest age, it is strongly associated with moist environments. Throughout its range, the PACIFIC WATER SHREW appears to be rare. There are no actual estimates of populations, but in most small-mammal studies, it represented less than 1 per cent of the small mammals captured. In 1992, Gustavo Zuleta and Carlos Galindo-Leal conducted an intensive survey of this species in southwestern BC using results from more than 19,000 trap nights in 55 locations. They captured 999 small mammals, but only three were PACIFIC WATER SHREWS, taken at three widely separated locations. Similar low numbers were reported in the study of Dale Seip and Jean-Pierre Savard. From 1989 to 1990 they sampled 17,315 pitfall trap nights covering 22 sites in the Capilano, Seymour and Coquitlam watersheds in the North Shore mountains. A total of five PACIFIC WATER SHREWS were captured at two of the 22 sites.

Semi-aquatic, PACIFIC WATER SHREW can swim underwater or on the surface, propelling itself with alternating movements of the hind legs. Its waterproof pelage keeps it warm in water, and air trapped in the fur provides buoyancy. There are reports of it running for several seconds across the surface of water.

The diet is rather specialized with aquatic insects such as water beetles and nymphal stages of stoneflies, mayflies and alderflies accounting for 25 per cent of the prey types eaten. Other food includes crane flies, carabid beetles, spiders, harvestmen, centipedes, earthworms, slugs and small terrestrial snails. The only information for BC is a historical museum specimen captured near Abbotsford with stomach contents of "water beetles, worms, insects." Much of what we know about the feeding behaviour of this species comes from observations of several captive individuals. They fed readily on earthworms, sowbugs, spiders and termites, but carabid beetles, dermestid beetles and crayfish were ignored. PACIFIC WATER SHREWS attacked earthworms ferociously with a series of swift bites along their bodies. Some earthworms were subdued by bites and then stored in the cage for later consumption. When hunting underwater, they seized

Water beetle prey.

an animal in their jaws and carried it to land before eating. PACIFIC WATER SHREW presumably uses its sensitive vibrissae to detect water movements and underwater sniffing to locate its prey, similar to the AMERICAN WATER SHREW (page 151).

A ball-like nest of a PACIFIC WATER SHREW was found one year on June 18 inside a woodpecker cavity in a fallen RED ALDER tree in Bear Creek, Surrey. The nest was constructed of dried grass and moss with soft strips of PAPER BIRCH. Six nearly naked young were in the nest. No other breeding data exist for the PACIFIC WATER SHREW in BC. In other parts of its range, the breeding season extends from late January to late August, with most young born in March. Litter size is usually three or four. The number of litters produced by females has not been documented. Males do not breed in their first summer.

No studies have been done on predation, but owls and fish are the most likely predators of this shrew. In 1999, Glenn Ryder discovered a PACIFIC WATER SHREW skull in the pellet of a BARN OWL found near Burns Bog. Accidental captures in fish traps may be a significant source of mortality. From 2000 to 2011, 10 voucher specimens of PACIFIC WATER SHREW deposited with the Royal BC Museum were fatalities found in submerged minnow traps used in fish surveys. Water shrews swim into the traps but are unable to escape and drown in the trap.

CONSERVATION STATUS

COSEWIC assessed this species as endangered in Canada, and it is listed under the Species at Risk Act. It is on the BC Red List. Rarity, specialized habitat requirements of riparian or wetland areas and a small distributional area in a heavily populated region undergoing rapid development are factors for its designation. Habitat loss from urban and industrial development, land clearing for agriculture, and forest harvesting are major concerns. A national recovery plan released in 2014 focuses on habitat protection and restoration, with 23 parcels of land within the known range designated as critical habitat. There are no data on historical or current population trends, but habitat loss has been extensive. As an example, Allan Brooks in 1902 described the PACIFIC WATER SHREW as "fairly common" in the Chilliwack valley, and from 1895 to 1897 he collected 41 specimens from the Sumas Prairie area that he sent to various natural history museums. In the 1920s, Sumas Lake and associated wetlands of the Sumas Prairie were drained and converted to agricultural land, with PACIFIC WATER SHREW habitat in this area now reduced to narrow riparian strips bordering dykes and canals.

SYSTEMATICS AND TAXONOMY

PACIFIC WATER SHREW is a member of the *Sorex vagrans* group, a complex of North American shrews that includes AMERICAN WATER SHEW, EASTERN WATER SHREW, WESTERN WATER SHREW, DUSKY SHREW, FOG SHREW, ORNATE SHREW, PACIFIC SHREW and VAGRANT SHREW. A recent molecular genetics study demonstrated that WESTERN WATER SHREW is the closest relative to PACIFIC WATER SHREW. Estimates based on the molecular clock hypothesis using the species's DNA sequences suggest that the two species diverged from a common ancestor some 410,000 years ago.

Three subspecies are recognized; one occurs in BC. There has been no modern morphological study or genetic analysis assessing the validity of the subspecies.

- *Sorex bendirii bendirii* (Merriam). A small race with dark ventral fur, widely distributed from northern California, the Cascade Mountains of Oregon, and western Washington (excluding the Olympic Mountains) to southwestern BC. The mean body mass of the BC sample is noticeably smaller than the mean body mass values given for the Oregon population, suggesting a south-to-north trend of decreasing size.

REMARKS

Because of its endangered status, BC government agencies have directed more attention at the PACIFIC WATER SHREW than any other shrew species. Recent surveys for environmental assessments yielded new information on habitat and distribution. But there are aspects of this shrew's biology that require dedicated research studies. What we know about its reproduction and food habits in BC, for example, is limited to a few observations recorded on historical museum specimen labels. Another topic warranting research is the degree of overlap in distribution among PACIFIC WATER SHREW and WESTERN WATER SHREW and possible competition. Both shrews were captured at Haney and Cultus Lake in the Fraser River valley. In the North Shore mountains, their distributions likely overlap at mid-elevations.

The species name *bendirii* is an eponym for Charles Bendire, a soldier in the US Army with a keen interest in natural history who collected the type specimen in 1882 while based at Fort Klamath, Oregon.

REFERENCES

Brooks (1902); COSEWIC (2006); Environment Canada (2014); Hope et al. (2014); Jackson (1951); Maser and Franklin (1974); McComb, McGarigal and Anthony (1993); Pattie (1969, 1973); Ryckman (2020); Ryder and Campbell (2007); Seip and Savard (1991); Verts and Carraway (1998); Whitaker and Maser (1976); Zuleta and Galindo-Leal (1994).

Cinereus Shrew

Sorex cinereus

Other common name: Masked Shrew

DESCRIPTION

Next to the WESTERN PYGMY SHREW and PREBLE'S SHREW, the CINEREUS SHREW is our smallest shrew. Its fur colour varies geographically. Interior populations are pale brown in dorsal pelage with greyish-white undersides; some individuals are faintly tricoloured with their sides paler than their backs. Coastal populations tend to be darker brown or grey. The winter pelage is more greyish than the summer pelage. The tail is pale underneath but not strongly bicoloured and has a tuft of darker hairs at the tip. The third, fourth and fifth toes on the hind foot have six pairs of toe pads.

Cranial/dental traits: 32 teeth; incisors 1/1, unicuspids 5/1, premolars 1/1, molars 3/3; upper incisor with a small to large medial tine situated from above to well below the upper edge of pigment; third upper unicuspid usually taller or equal to the fourth; a strongly pigmented ridge on the upper unicuspids extends to the cingulum and is not separated by a longitudinal groove; lower incisor has three denticles with continuous pigmentation on labial side extending to either the first, second or third denticle, with pigment on the lingual side extending to the second or third denticle; small to large postmandibular foramen in one or both dentaries observed in about one-third of the individuals examined.

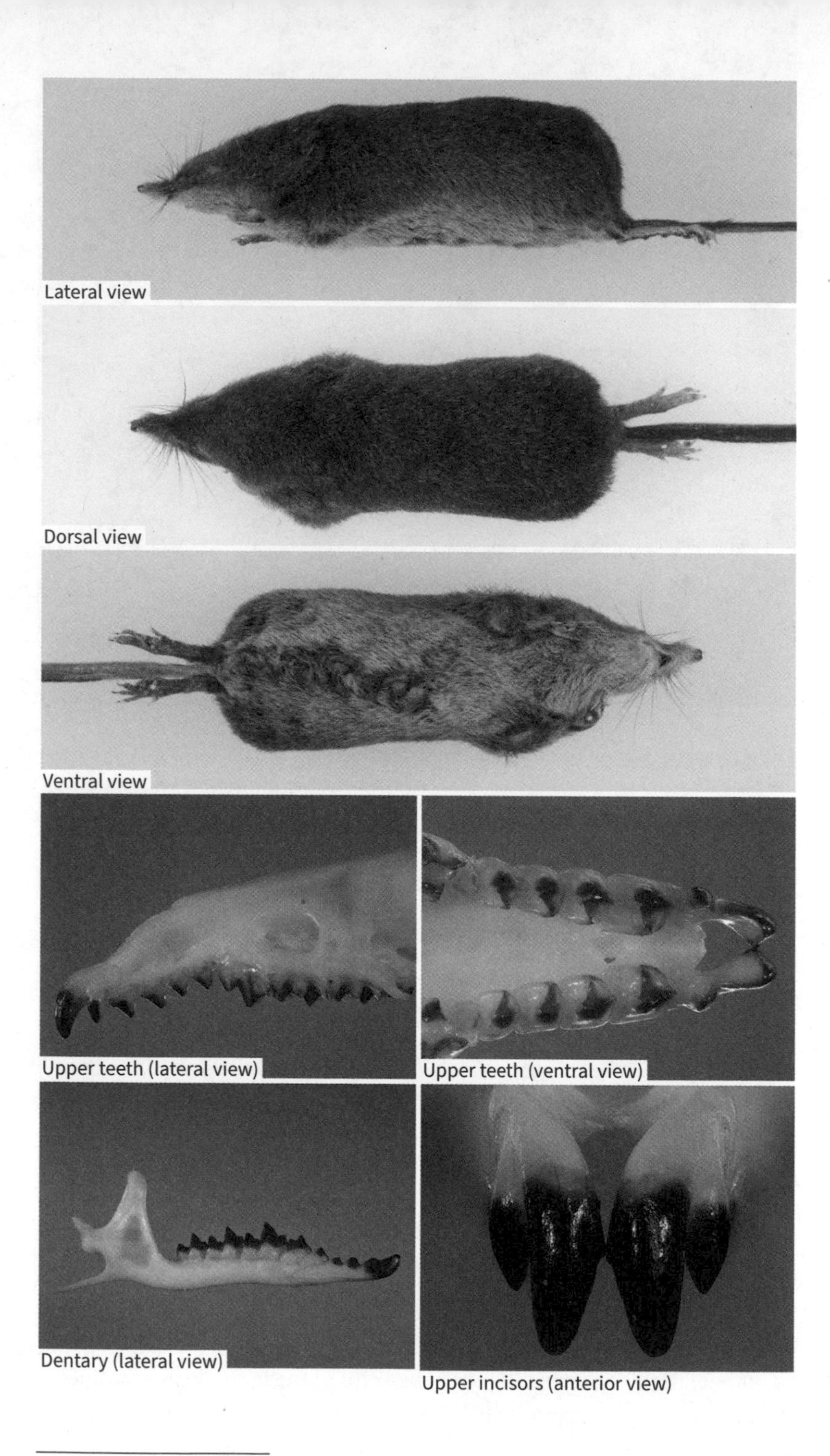

Upper incisors (anterior view)

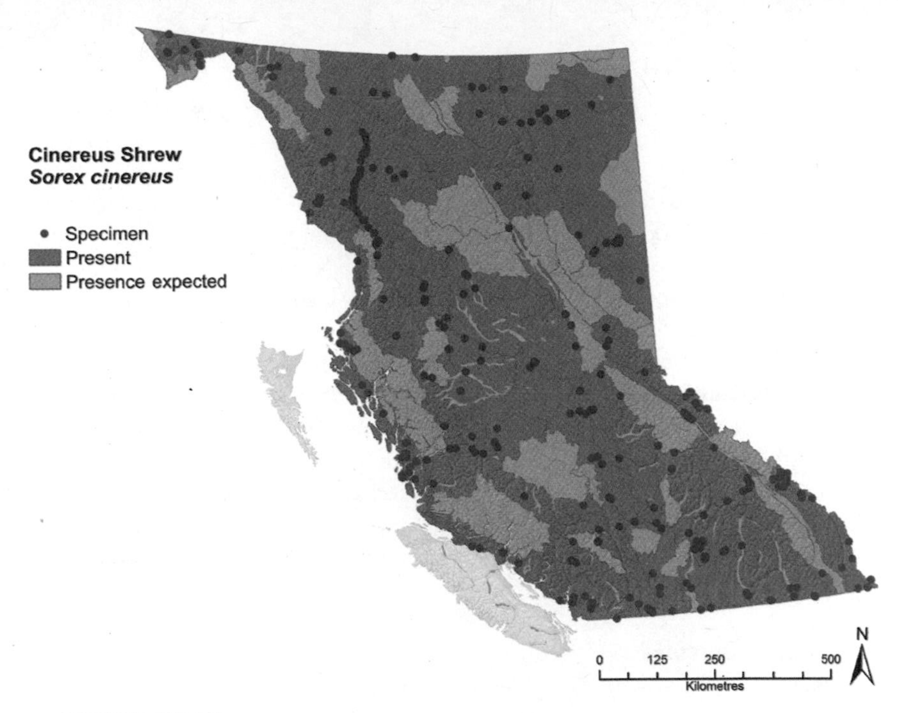

DISTRIBUTION

CINEREUS SHREW has the largest distributional area of any North American shrew; it is found across the northern United States, throughout most of Canada and in Alaska. In BC, it inhabits the entire mainland and some islands on the central and north coast: Athlone (formerly Smyth) and Townsend* in the Bardswell Group, Campbell, Hunter, Kaien, King, McCauley, Pitt, Princess Royal, Ruth, Spider, several islands in the Tribal Island group and Yeo. It is notably absent from Haida Gwaii, Vancouver Island and islands in Johnstone Strait and the Strait of Georgia.

MEASUREMENTS

	Mean	Range	Sample size
Total length:	98 mm	76–121 mm	*n* = 612
Tail vertebrae:	42 mm	30–53 mm	*n* = 608
Hind foot:	12 mm	8–14 mm	*n* = 607
Mass—females:	3.8 g	2.0–7.0 g	*n* = 67
Mass—males:	4.2 g	2.0–7.0 g	*n* = 84

* This is the island name given on specimen tags and in the paper by McCabe and McTaggart Cowan (1945), but we could not find it on any map, and which island in the Bardswell Group is Townsend Island is unknown.

MORPHOLOGICALLY SIMILAR SPECIES

CINEREUS SHREW co-occurs with 13 BC shrew species. ARCTIC SHREW, TUNDRA SHREW, AMERICAN WATER SHREW, WESTERN WATER SHREW and PACIFIC WATER SHREW are easily distinguished from external features. Other species require examination of cranial/dental traits. DUSKY SHREW, PACIFIC SHREW, TROWBRIDGE'S SHREW and VAGRANT SHREW have a third upper unicuspid smaller than the fourth. WESTERN PYGMY SHREW has minute third and hidden fifth upper unicuspids with only three unicuspids readily visible in side view; MERRIAM'S SHREW lacks a medial tine on its upper incisor and has very shallow spaces between the denticles on its lower incisor. Most likely to cause misidentification are PREBLE'S SHREW and OLYMPIC SHREW. PREBLE'S SHREW is distinguished from CINEREUS SHREW by smaller cranial and dental measurements: skull length less than 14.5 millimetres, length of mandible less than 6.5 millimetres. OLYMPIC SHREW either lacks a medial tine or has a minute medial tine on its upper incisor that is positioned at or above the pigment edge. It also has a narrower first upper unicuspid (width less than 0.56 millimetres) than that of CINEREUS SHREW.

NATURAL HISTORY

CINEREUS SHREW is found in all of the province's ecoprovinces with an elevational range ranging from sea level to 2,288 metres in the southern Rocky Mountains (Mount Assiniboine Provincial Park) and 2,286 metres elevation in the southern Coast Mountains (Cayoosh Range). It is associated with many habitats: open and closed forests, open meadows, avalanche slopes, riverbanks and lakeshores, TAMARACK–BLACK SPRUCE bogs and DWARF BIRCH–willow thickets. Habitat selection correlates strongly with moisture with a preference for mesic forest and wetlands. In dry forests near Kamloops, significantly higher captures of CINEREUS SHREW were in TREMBLING ASPEN stands than in the drier DOUGLAS-FIR or mixed-forest habitats. Habitats disturbed by fire or logging are readily exploited. Walt Klenner found this shrew in all his forest study plots in the Thompson Plateau, from uncut stands to recent clearcuts.

CINEREUS SHREW is often the dominant shrew species in many habitats in the BC Interior. At Opax Mountain near Kamloops, CINEREUS SHREW was the most commonly captured of the four shrew species present. However, in coastal forests PACIFIC SHREW is typically more abundant. Some of the highest CINEREUS SHREW populations occur in habitats with standing water. In Manitoba, population densities of 5 to 22 per hectare were recorded in bog habitats. Nevertheless, abundance varies extensively among habitat types, and pronounced population fluctuations occur from year to year that may be related to variations in the abundance of prey. Home-range estimates for five CINEREUS SHREWS captured by Vanessa Craig in her 60-to-80-year-old forest-study areas in the Vancouver watershed ranged from 307 to 1,840 square metres. Evidently this species uses vole runways. James Munro captured 54 CINEREUS SHREWS in vole runways in meadows at Sinkut Mountain in the Nechako Plateau.

Activity bouts are for short periods (about two minutes) throughout a 24-hour period. Its peak activity is after dark, and activity is greatly increased when there is a nighttime rainfall. Most activity is associated with feeding; captive CINEREUS SHREWS, observed for a week, fed about every 13 minutes.

Dietary data are available from a number of studies done in eastern North America. Prey identified in the stomachs of captured shrews included insect larvae, ants, beetles, crickets, grasshoppers, spiders, harvestmen, centipedes, slugs, snails and fungi. Seeds may also be consumed in winter. The diet is flexible with considerable geographic variation in diet. A study in Michigan revealed that ants were the major prey, representing 50 per cent of the food items; in New Brunswick, insect larvae were recorded as the predominant prey type. CINEREUS SHREWS associated with TAMARACK bogs in Manitoba fed almost entirely on LARCH SAWFLY larvae in late summer. A population living on a small island in Nova Scotia hunted mainly in the intertidal zone where they fed on kelp flies and marine amphipods.

In Manitoba, two CINEREUS SHREWS were observed hunting butterflies during the day. Darting quickly toward their prey, the shrews leaped into the air to pounce on the butterflies before they could take flight. They ate only the bodies of the butterflies and discarded the wings. CINEREUS SHREWS were observed raiding a BLUE-HEADED VIREO nest 1.7 metres above ground suggesting that some will hunt in shrubs and trees.

Data from BC museum specimens and field studies indicate that females breed from May to September. In eastern North America, the breeding season is generally from April to October, and in populations with abundant food resources, it may extend into November. Average embryo counts reported for various populations across North America range from five to eight. Females can produce at least two litters in a breeding season. Although this is uncommon, males and females may breed in their first summer. Newborn CINEREUS SHREWS weigh 0.2 to 0.3 grams and are only 12 to 14 millimetres long. They grow quickly, attaining adult size in 20 to 27 days when they leave the nest. Less than half the young will survive beyond five months. The maximum life span is about 15 months, although a few individuals may reach two years of age. Predators include owls and small carnivores. An single owl pellet found in the Baker Creek valley near Quesnel contained seven CINEREUS SHREW skulls. In southwestern Alaska, several CINEREUS SHREWS were found in the stomachs of two ARCTIC GRAYLINGS. The shrews may have accidentally entered the water while foraging along the edge of salmon streams.

CONSERVATION STATUS

Occupying a wide range of habitats and a large distributional area, CINEREUS SHREW is considered secure in BC.

SYSTEMATICS AND TAXONOMY

CINEREUS SHREW is a member of the *Sorex cinereus* group that comprises 13 related species found in North America and Siberia. A study of mitochondrial and nuclear DNA confirmed that the group splits into two major genetic lineages—Beringian and Southern. CINEREUS SHREW is a member of the Southern lineage. Four clades of CINEREUS SHREW were evident in the phylogenetic tree with coastal and mainland BC populations all aligning with the Cinereus west clade. Six subspecies are recognized in CINEREUS SHREW with two occurring in BC. But they are inconsistent with the four genetic clades.

- *Sorex cinereus cinereus* Kerr. Distributed across much of Canada and the entire province east of the coastal mountain ranges. Measurements: total length 97 mm (76–121 mm), *n* = 463; tail vertebrae 41 mm (30–53 mm), *n* = 461; hind foot 12 mm (8–14 mm), *n* = 458; mass 3.9 g (2–7 g), *n* = 158.
- *Sorex cinereus streatori* Merriam. Inhabits the Pacific coast from Washington to Alaska. In BC, it is found on western slopes of the Coastal mountain ranges, coastal lowlands and some islands on the central and north coast. Has a slightly larger body size and darker summer pelage than *Sorex cinereus cinereus*. Measurements: total length 102 mm (86–120 mm), *n* = 149; tail vertebrae 44 mm (32–53 mm), *n* = 147; hind foot 12 mm (10–14 mm), *n* = 149; mass 3.9 g (2.5–5.0 g), *n* = 32.

REMARKS

It is curious that few coastal islands are occupied by the CINEREUS SHREW, given that it occupies the entire adjacent coastal mainland. Ian McTaggart Cowan reported that this species was less abundant than the ubiquitous PACIFIC SHREW on the north coast islands. Low populations and a small body size may limit this shrew's ability to colonize and survive on these islands. It is also possible that CINEREUS SHREW is unable to outcompete the larger PACIFIC SHREW on some islands particularly those with limited food resources.

The species name *cinereus* means "ash grey," a reference to this shrew's pelage colour.

REFERENCES

Buckner (1964, 1966, 1970); Craig (1995); Forsyth (1976); Hope et al. (2012); Horvath (1965); Huggard and Klenner (1998); McTaggart Cowan (1941); Moore and Kenagy (2004); Munro (1947, 1955); Oaten and Larsen (2008); Stewart, Herman and Teferi (1989); Teferi, Herman and Stewart (1992); Whitaker (2004).

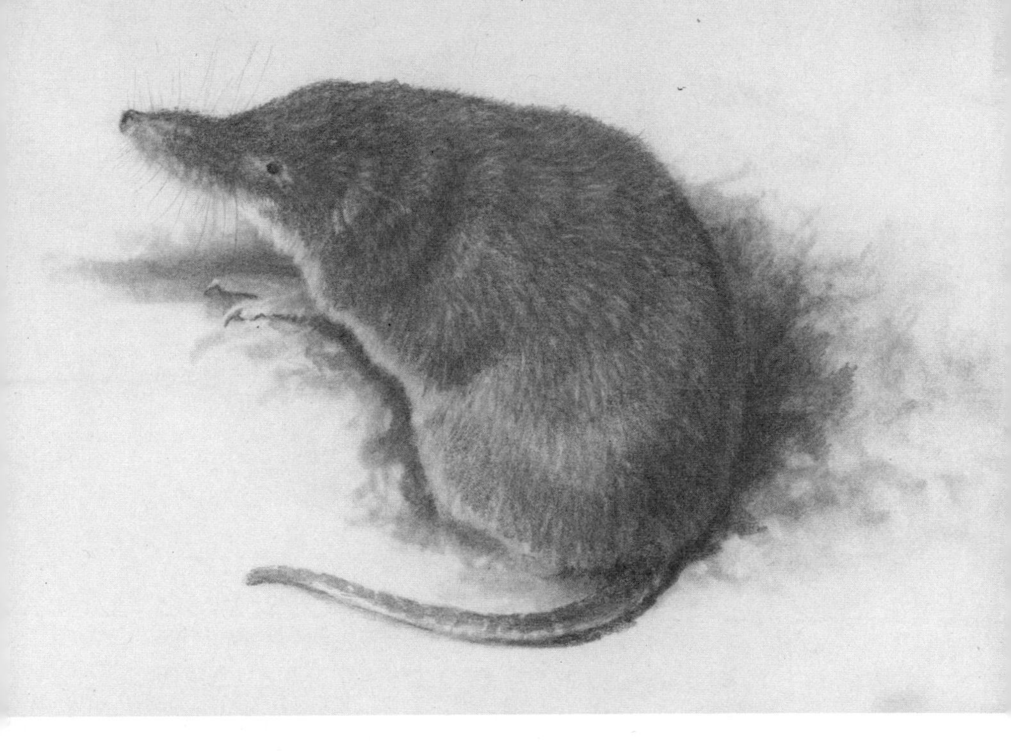

Western Pygmy Shrew *Sorex eximius*

DESCRIPTION

WESTERN PYGMY SHREW is the smallest shrew in BC. Its dorsal fur is dull greyish brown; the undersides are light grey or brown. The winter pelage is more grey than brown and appears slightly tricoloured. The tail is weakly bicoloured and relatively short, and usually makes up less than 40 per cent of the total length. The head appears shorter than those of other BC shrews.

Cranial/dental traits: 32 teeth; incisors 1/1, unicuspids 5/1, premolars 1/1, molars 3/3; teeth darkly pigmented; upper incisor with a large, long medial tine in the pigmented area on the face of the incisor separated from it by a deep groove; upper toothrow appears crowded; small disc-like third unicuspid and fifth unicuspid not readily visible in side view; first upper unicuspid with a pigmented ridge that curves posteriorly and is separated from the cingulum by a groove; dentary short and robust; lower incisor has three denticles with the third denticle very small, and pigment area on the labial side covering all three denticles, with lingual pigment extending to the second denticle; dentaries lack a postmandibular foramen.

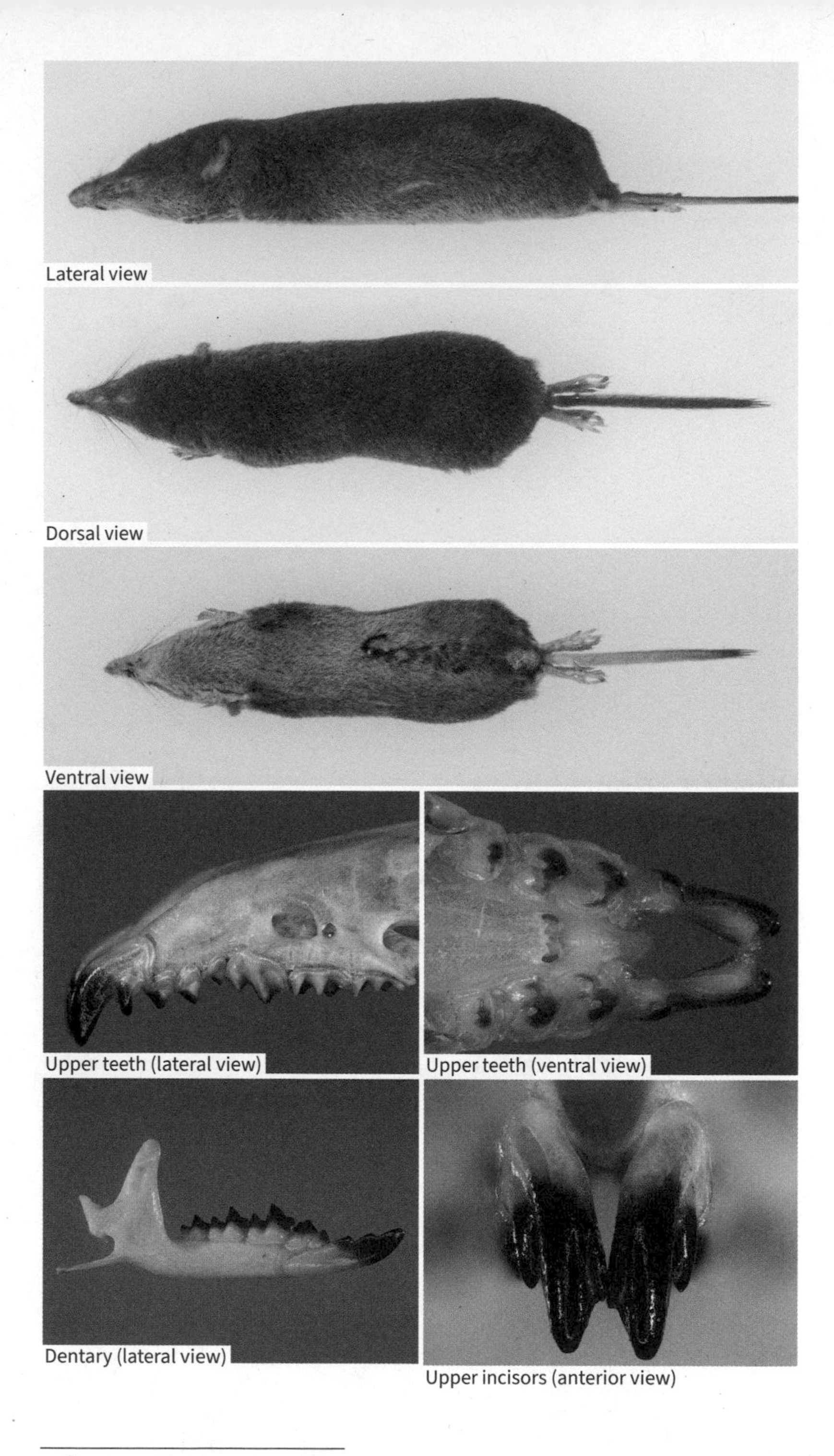
Lateral view

Dorsal view

Ventral view

Upper teeth (lateral view)

Upper teeth (ventral view)

Dentary (lateral view)

Upper incisors (anterior view)

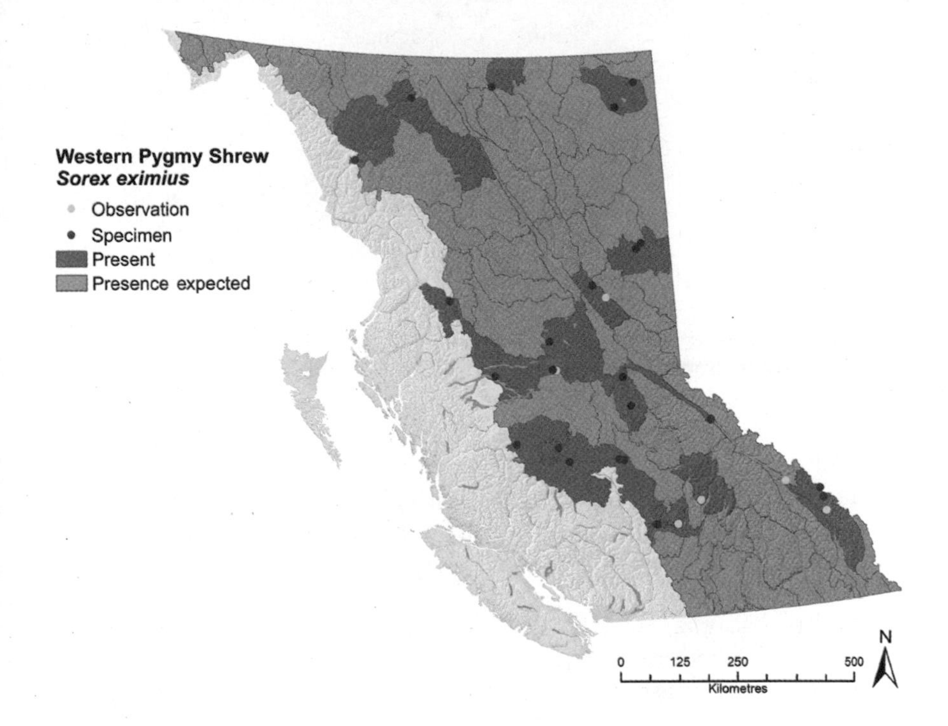

DISTRIBUTION

WESTERN PYGMY SHREW ranges across western North America from Alaska south to the northern and central Rocky Mountains of the United States. The map in the study by Andrew Hope and colleagues shows it extending as far east as Manitoba and North Dakota. Because it can only be identified by genetics (see Systematics and Taxonomy below), large geographic gaps in genetic sampling particularly from the Prairies prevent a firm determination of the eastern limits of its range. In BC, it occurs east of the Cascade and Coast Mountains from the Yukon border as far south as the Kamloops area and the southern Rocky Mountains (Kootenay National Park). Although it is present in dry forests on Opax Mountain near Kamloops and occupies shrub-steppe habitats in Idaho and Montana, there are no known occurrences from the Okanagan or Similkameen valleys. The sparseness and wide scattering of occurrences in the province are probably due to inadequate sampling.

MEASUREMENTS

Measurements are based on specimens from BC, southern Yukon and Washington State.

	Mean	Range	Sample size
Total length:	83 mm	73–96 mm	$n=29$
Tail vertebrae:	28 mm	22–33 mm	$n=29$
Hind foot:	10 mm	6–12 mm	$n=29$
Mass:	3.4 g	2.0–5.5 g	$n=41$

MORPHOLOGICALLY SIMILAR SPECIES

BC species most likely to be confused with WESTERN PYGMY SHREW are CINEREUS SHREW, PREBLE'S SHREW and VAGRANT SHREW. In external features, WESTERN PYGMY SHREW is identifiable by its relatively shorter tail (less than 40 per cent of the total length) and greyish pelage. The presence of only three conspicuous upper unicuspid teeth when the skull is viewed laterally is a diagnostic trait that distinguishes WESTERN PYGMY SHREW from all other BC shrews. The short skull and dentary (skull length less than 15.0 millimetres, mandible length less than 6.5 millimetres) will discriminate it from all but PREBLE'S SHREW.

NATURAL HISTORY

WESTERN PYGMY SHREW can be found in forested habitats, wetlands and bogs. Ideal conditions appear to be boreal habitats with a mixture of wet and dry soils. It will occupy disturbed habitats such as recently logged or burned forest, flooded areas and cultivated land. Information from museum specimens and a number of small-mammal surveys indicates a diversity of habitat use in the province. Records are from tall grass bordering lakes; TREMBLING ASPEN, ENGELMANN SPRUCE–SUBALPINE FIR, and aspen–LODGEPOLE PINE forests; dry DOUGLAS-FIR forests of various harvesting stages; a subalpine burn; stump piles in recently cleared land; open willow thickets; a forest bordering a reservoir; and TAMARACK–BLACK SPRUCE bogs. James Munro reported a capture in a vole runway in meadows at Sinkut Mountain in the Nechako Plateau. In Washington and Idaho, WESTERN PYGMY SHREWS were captured in second-growth forests dominated by LODGEPOLE PINE, PONDEROSA PINE and DOUGLAS-FIR, and in subalpine forests dominated by SUBALPINE FIR and GRAND FIR. It has also been found in sagebrush-steppe habitat in Idaho and Montana. Its elevational range in BC extends from 300 to 1,647 metres in the Rocky Mountains (Kootenay National Park).

There are no estimates of home-range size or population density for WESTERN PYGMY SHREW. It is generally described as uncommon in most communities, but its abundance may be underestimated because of ineffective sampling methods. In BC, the dominant shrews in communities where WESTERN PYGMY SHREW has been found are VAGRANT SHREW, DUSKY SHREW or CINEREUS SHREW. David Huggard and Walt Klenner captured surprisingly high numbers of WESTERN PYGMY SHREWS (8 per cent of total mammal captures) in their pitfall trap sites at Opax Mountain near Kamloops. Shrews most commonly captured in their study area were CINEREUS SHREW and DUSKY SHREW, but WESTERN PYGMY SHREWS were more common than VAGRANT SHREWS.

No descriptions of food habits are available for this species. The closely related EASTERN PYGMY SHREWS seem to prefer small invertebrates less than five millimetres long. Larger prey, such as earthworms, snails and slugs, are not eaten; they may be too large to handle. Breeding data are scanty. In BC, nursing females have been found in June and a pregnant female with nine well-developed embryos was captured on June 21. Predators include owls. In northern Alberta, a WESTERN PYGMY SHREW was found in the nest box of a NORTHERN SAW-WHET OWL and a pellet of a GREAT HORNED OWL. A WESTERN PYGMY SHREW was recovered in the stomach of an ARCTIC GRAYLING in southern Yukon.

CONSERVATION STATUS

Considered secure in BC, it is found in a wide spectrum of habitat types. Its alleged rarity may reflect sampling bias more than low population densities.

SYSTEMATICS AND TAXONOMY

Until recently, North American pygmy shrews were classified as a single species *Sorex hoyi* with five subspecies recognized on the basis of morphology. In 2003, a preliminary assessment of mitochondrial DNA revealed two genetic lineages that were suggestive of two species. A more comprehensive genetic study in 2019 with broader geographic coverage confirmed two lineages consistent with distinct eastern (EASTERN PYGMY SHREW) and western (WESTERN PYGMY SHREW) species. They diverged from a common ancestor 370,000 years ago. They show no unequivocal differences in their morphology and can only be identified from genetics.

We adopted the recommendations of the recent genetic research and treat the BC populations as WESTERN PYGMY SHREWS. No genetic samples from BC were actually analyzed in the 2019 study, but samples from the Rocky Mountains of Alberta and southeastern Yukon near the BC border were genotyped as WESTERN PYGMY SHREWS.

Within the WESTERN PYGMY SHREW, two clades are evident from DNA analysis—western and southwestern. BC populations are associated with the western clade. Separated geographically and differing in their morphology, the clades are consistent with their recognition as two subspecies.

- *Sorex eximius eximius* (Osgood). Occupies a wide range across northwestern North America and the northern Rocky Mountains of the United States. The eastern limits of its range have not been determined but are likely eastern Manitoba and North Dakota. There is a general trend of decreasing body and cranial/dental size from north to south with the largest individuals in Alaska.

REMARKS

With the recent taxonomic revision, it appears that much of the published literature on the biology of North American pygmy shrews may apply to EASTERN PYGMY SHREW. For this account we only used studies from western North America. It is noteworthy that although the range map in the 2019 study shows WESTERN PYGMY SHREWS occurring in BC, Alberta, Saskatchewan and most of Manitoba, only three genetic samples from two locations in the Alberta Rocky Mountains were available from this vast area. More genetic samples from across central and western Canada are required to establish the range limits of the two pygmy shrew species.

The species name *eximius* is from Latin, meaning "uncommon" or "extraordinary."

REFERENCES

Diersing (1980); Foresman (1999); Hope et al. (2020); Huggard and Klenner (1998); Jung et al. (2011); Munro (1955); Schowalter (2002); Stewart et al. (2003); Stinson and Reichel (1985).

Merriam's Shrew *Sorex merriami*

DESCRIPTION

MERRIAM'S SHREW is a small shrew with a pale, drab greyish-brown pelage dorsally and nearly white ventral fur. The tail is distinctly bicoloured pale brown dorsally and nearly white underneath with a white tip in some individuals. In Washington State, this species has distinct summer and winter pelages, with the moult to summer pelage beginning in April and the moult to winter pelage beginning in September. The winter pelage is more grey dorsally and paler, nearly white ventrally.

Cranial/dental traits: 32 teeth; incisors 1/1, unicuspids 5/1, premolars 1/1, molars 3/3; teeth are strongly pigmented, tall and robust; upper incisor lacks a medial tine; fifth upper unicuspid partly hidden by first molar in side view; in animals with unworn teeth, distal tips of the upper incisors converge; third upper unicuspid usually taller than the fourth; cusps of upper unicuspids with a strongly pigmented medial ridge that does not reach the cingulum; lower incisor has three weakly developed denticles, with continuous pigmentation on the labial side extending to third denticle, and lingual pigment reaching the second or third denticle; small to large postmandibular foramen present in both dentaries.

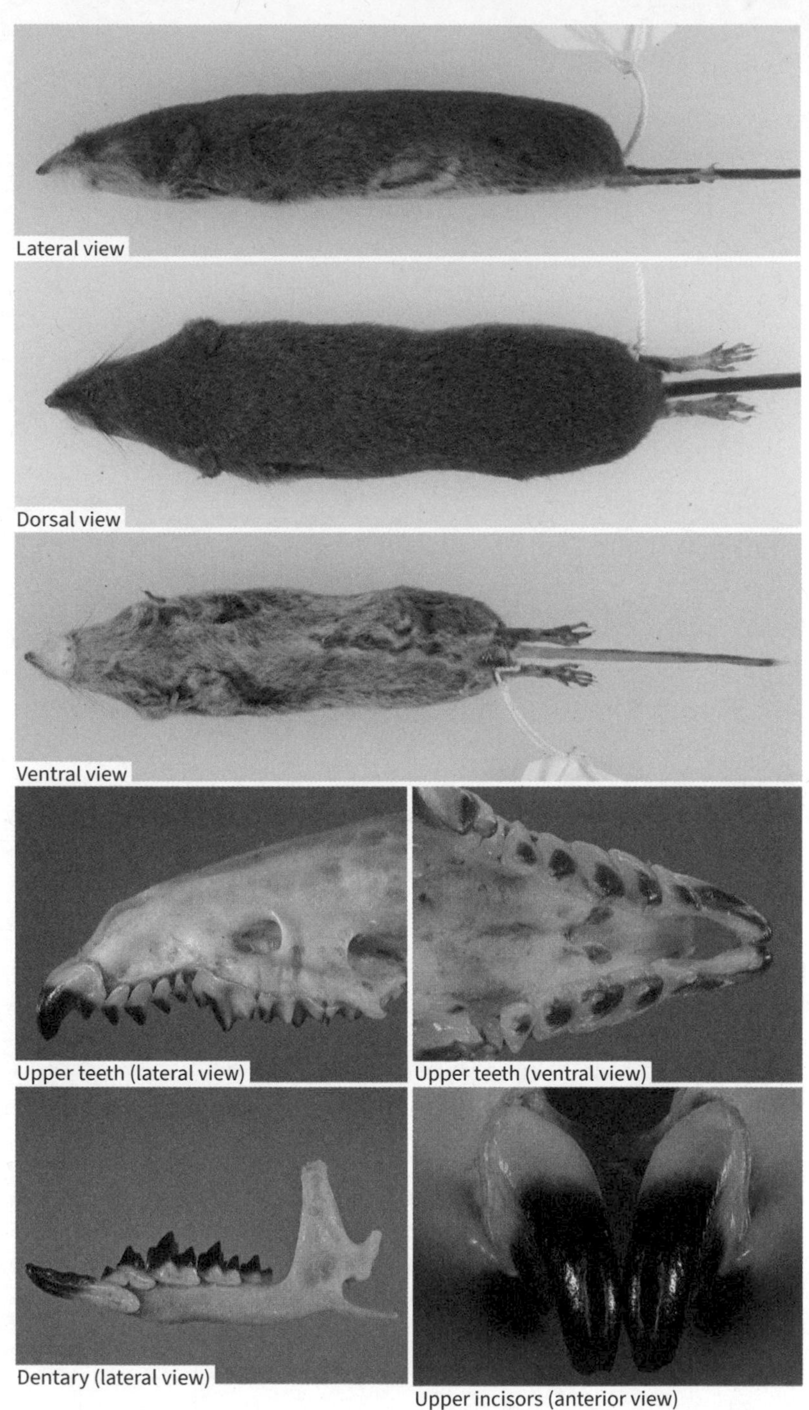
Lateral view
Dorsal view
Ventral view
Upper teeth (lateral view)
Upper teeth (ventral view)
Dentary (lateral view)
Upper incisors (anterior view)

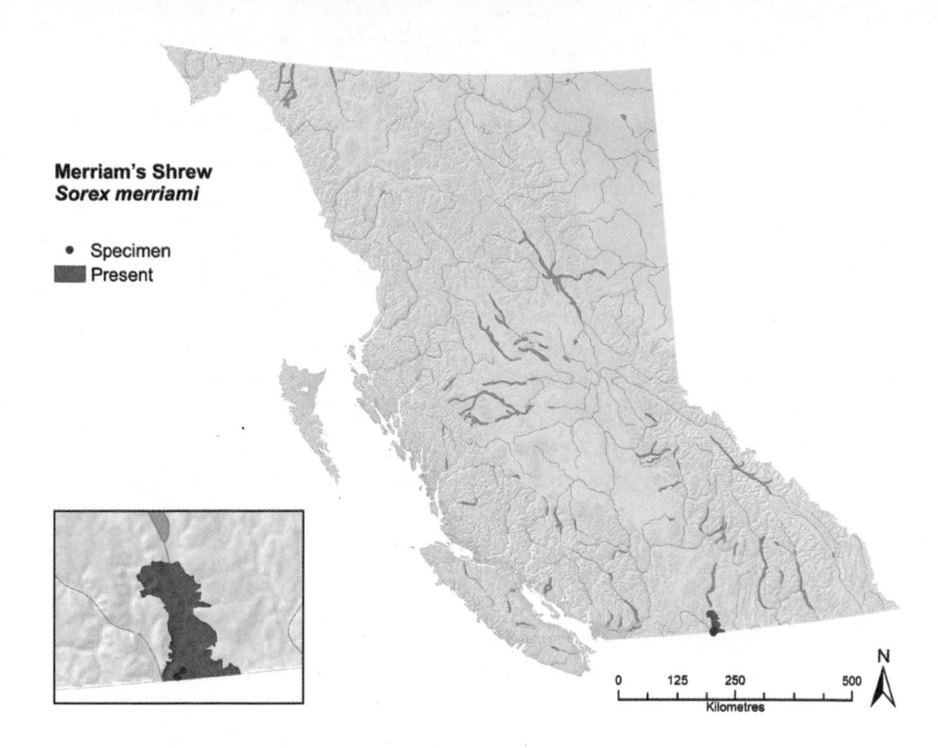

DISTRIBUTION

MERRIAM'S SHREW is patchily distributed across the western United States east of the Sierra Nevada and Cascade Mountains in the Great Basin, Columbia Plateau and Great Plains. It is known from only two capture locations in the southern Okanagan region of BC, both south of Richter Pass near the Washington border—Kilpoola Lake and the Sparrow Grasslands property of the Nature Conservancy of Canada. Given its distribution in the Columbia Basin of Washington, it would be expected to occupy suitable habitats in the southern Okanagan and Similkameen valleys.

MEASUREMENTS

Measurements are based on specimens from Washington State.

	Mean	Range	Sample size
Total length:	96 mm	85–106 mm	$n=23$
Tail vertebrae:	37 mm	33–40 mm	$n=23$
Hind foot:	12 mm	11–15 mm	$n=23$
Mass:	4.2 g	2.8–5.9 g	$n=22$

MORPHOLOGICALLY SIMILAR SPECIES

In BC, CINEREUS SHREW, PREBLE'S SHREW and VAGRANT SHREW are the species most likely to co-occur with MERRIAM'S SHREW and cause identification problems. Distinguishing them from MERRIAM'S SHREW by external features is not possible, and identification requires cranial/dental traits or genetic samples. The presence of prominent medial tines on the upper incisors and deep spaces between the denticles on the lower incisor readily identify PREBLE'S SHREW and CINEREUS SHREW. They can also be discriminated by some smaller cranial/dental measurements. PREBLE'S SHREW has a skull length less than 14.5 millimetres; CINEREUS SHREW has a skull length less than 16.0 millimetres, and the width across the third upper molars is less than 4.0 millimetres. VAGRANT SHREW overlaps in cranial measurements but has a medial tine on its upper incisor, although in some individuals the tine is reduced in size and is pale, lacking strong pigmentation; its third unicuspid is smaller than the fourth, and the lower incisor has deep spacing between the denticles.

NATURAL HISTORY

Both BC captures were in grassland with BIG SAGEBRUSH at 827 to 918 metres elevation. Habitat at the capture sites was similar with sparse shrub cover (5–10 per cent) and more than 50 per cent grass cover with no tree cover. Common forbs were ARROWLEAF BALSAMROOT, YARROW and TIMBER MILK-VETCH. No other shrew species were captured at the two MERRIAM'S SHREW capture sites.

Capture site at Nature Conservancy of Canada lands, south of Richter Pass.

There are no population estimates for this shrew, but most studies describe it as rare. It is captured infrequently and in low abundance in the Columbia Basin of Washington. Sampling of 20 study sites in the south Okanagan with arthropod pitfall traps from 1993 to 1998 by Geoff Scudder yielded only 24 shrew captures of which one was MERRIAM'S SHREW, the first capture for Canada reported by Nagorsen and colleagues. In 2013, a survey employing 4,570 trap days in three grassland properties of the Nature Conservancy of Canada near the 1996 capture site produced only one MERRIAM'S SHREW.

No information on other aspects of this shrew's natural history in BC are available. We used findings from the Murray Johnson and Wesley Clanton study in central Washington for data on reproduction and diet. They observed pregnant females from April to July; three pregnant females had five to seven embryos. Enlarged testes were observed in male captures from March to early June. An analysis of stomach and intestinal contents revealed spiders, beetles, caterpillars, cave crickets and Ichneumon flies in prey remains. For its small body size, MERRIAM'S SHREW has a strong bite force and robust teeth that may be an adaptation for eating hard-bodied arthropods. The inward converging upper incisors with no medial tines may assist in the rapid piercing of prey. MERRIAM'S SHREW remains were recovered in the pellets of GREAT HORNED OWLS in Wyoming and MEXICAN SPOTTED OWLS in New Mexico.

CONSERVATION STATUS

This species is on the province's Red List; there has been no COSEWIC assessment. The indeterminate range extent in the province and lack of data on population trends prohibit a reliable status assessment. Rarity and the ongoing loss of shrub-steppe habitat in the south Okanagan make it a mammal of concern. The impacts of cattle grazing on this species have not been studied. The BC population appears to be isolated with the nearest known occurrences in Washington State about 170 kilometres south of the international border in the Columbia Basin. This distributional gap likely reflects the lack of shrew surveys in northern Washington.

SYSTEMATICS AND TAXONOMY

We could find no published information on DNA sequences of MERRIAM'S SHREW. Sarah George's phylogenetic study based on allozymes revealed that MERRIAM'S SHREW, TROWBRIDGE'S SHREW and ARIZONA SHREW are members of a distinct clade of North American shrews. This ancient lineage may have diverged from other North American shrews in the late Pliocene epoch several million years ago. No subspecies are recognized for MERRIAM'S SHREW.

REMARKS

The discovery of MERRIAM'S SHREW in BC can be attributed to Geoff Scudder, an entomologist who found it in 1996 in one of his terrestrial arthropod traps during his surveys in the south Okanagan valley. Fortunately, he deposited his incidental small-mammal fatalities as voucher specimens with the Royal BC Museum. It was a surprise when we identified this species in his collection, as it was not known to occur in Canada.

The species name *merriami* is an eponym for Clinton Hart Merriam, a prominent mammalogist who developed the US Biological Survey and its program of faunal surveys throughout North America.

REFERENCES

Armstrong and Jones (1971); Carraway and Verts (1994); Diersing and Hoffmeister (1977); George (1988); Gitzen, West and Trim (2001); Hudson and Bacon (1956); Johnson and Clanton (1954); Nagorsen et al. (2001); Tye, Geluso and Fugagli (2016).

Western Water Shrew *Sorex navigator*

Other common name: Navigator Shrew

DESCRIPTION

WESTERN WATER SHREW is a large distinctive shrew with dark-grey or black dorsal fur, although some appear dull brown; its ventral fur is silver grey, sometimes washed with brown. Scattered silvery-white hairs are also visible on some animals. The tail is distinctly bicoloured, dark above and white or grey below. The ventral surface of the tail for about the last centimetre from the tip has a raised fleshy ridge about one millimetre high fringed by stiff hairs. This forms a widening of the tail compared to other shrews and presumably functions to aid propulsion in water. The chin is pale white or grey. The hind feet are large with a fringe of stiff hairs about one millimetre long on the outer margins. Smaller stiff hairs occur on the front feet.

Cranial/dental traits: skull large with a rostrum that is not down-curved in side profile; 32 teeth; incisors 1/1, unicuspids 5/1, premolars 1/1, molars 3/3; upper incisor with small to medium-sized medial tine well within pigmented area; third upper unicuspid distinctly shorter than fourth; upper unicuspids with a weak to strongly pigmented ridge that extends to the cingulum; lower incisor has three denticles with continuous pigment on the labial side that covers all three denticles, with lingual pigment extending to the second or third denticle; tiny to large postmandibular foramen present in about half of the individuals we examined, usually present in only one dentary.

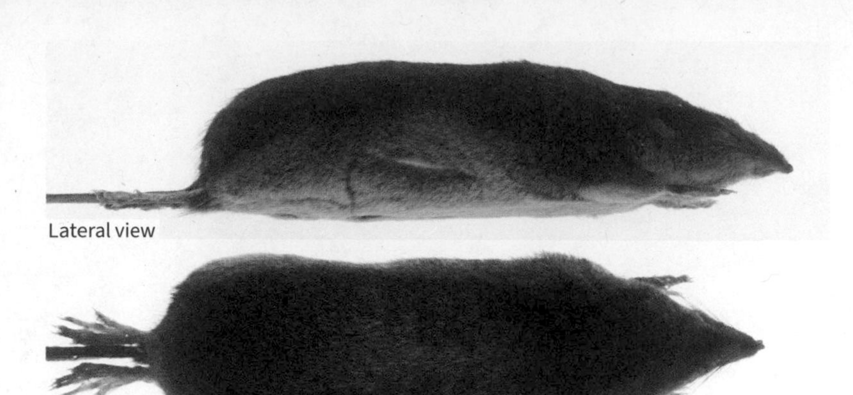

Lateral view

Dorsal view

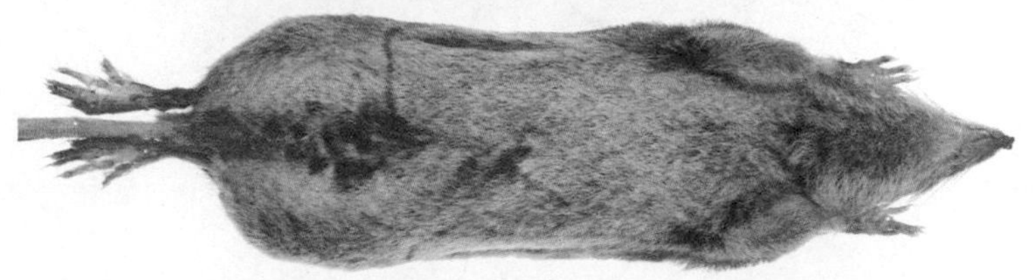

Ventral view of the subspecies *Sorex navigator navigator*

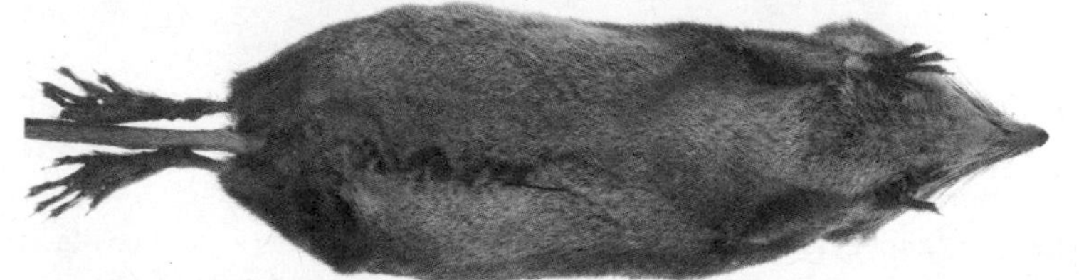

Ventral view of the subspecies *Sorex navigator brooksi*

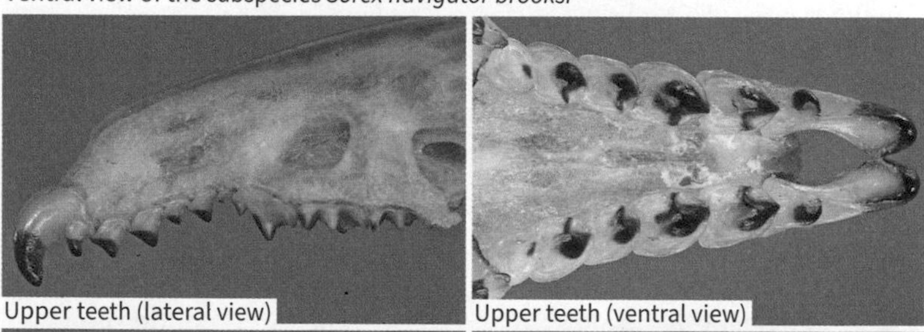

Upper teeth (lateral view)

Upper teeth (ventral view)

Dentary (lateral view)

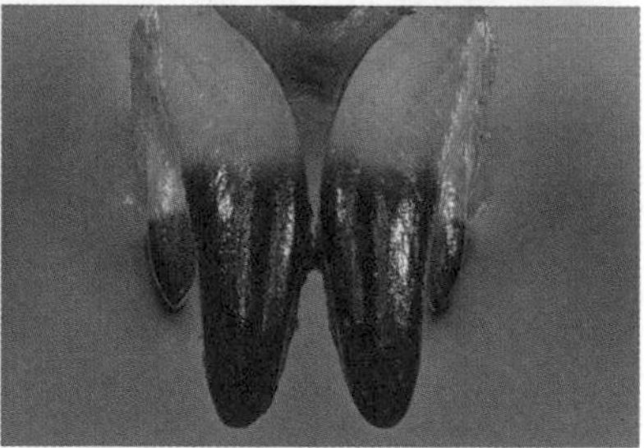

Upper incisors (anterior view)

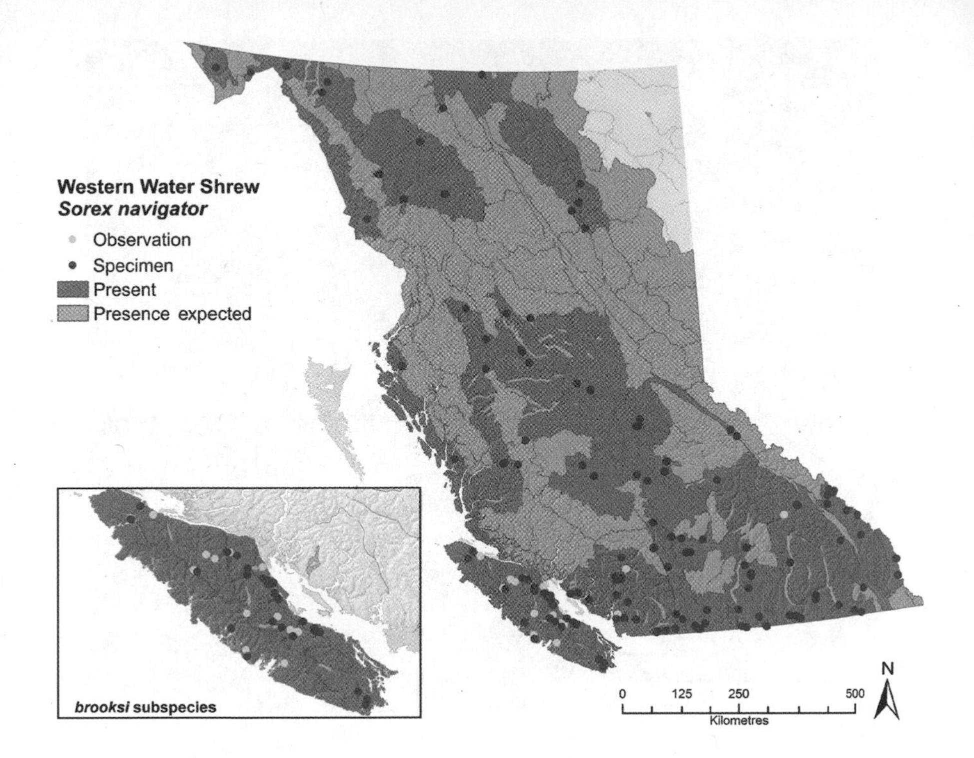

DISTRIBUTION

WESTERN WATER SHREW inhabits the western cordillera and northern coastal regions from Alaska and western Canada, south to New Mexico and Arizona in the United States. Eastern limits in Canada are the foothills of the Rocky Mountains. Its range includes the entire mainland of the province, except for the extreme lower Fraser River valley and northeastern BC where the eastern range limit is likely the foothills of the northern Rocky Mountains. A population on Vancouver Island is the only known coastal island population of WESTERN WATER SHREW in BC; however, it could inhabit large islands in the Great Bear Rainforest such as Princess Royal Island and Pitt Island.

MEASUREMENTS

	Mean	Range	Sample size
Total length:	150 mm	123–180 mm	*n* = 249
Tail vertebrae:	74 mm	59–88 mm	*n* = 248
Hind foot:	20 mm	16–22 mm	*n* = 249
Mass—females:	10.3 g	6.9–17.4 g	*n* = 37
Mass—males:	11.7 g	7.3–21.5 g	*n* = 37

Typical habitat on the BC mainland.

MORPHOLOGICALLY SIMILAR SPECIES

The only species that could cause misidentification are the two other water shrews found in the province—AMERICAN WATER SHREW and PACIFIC WATER SHREW. Although known occurrences of AMERICAN WATER SHREW and WESTERN WATER SHREW in northeastern BC are separated by a 200-kilometre gap, their distributions likely overlap along the foothills of the northern Rocky Mountains. Our research has shown that they are cryptic species indistinguishable using external traits and only identifiable by several cranial/dental measurements taken on prepared museum specimens (see the Identification Key to Cranial/Dental Traits, page 52) or by genetic sampling. PACIFIC WATER SHREW and WESTERN WATER SHREW ranges overlap in the North Shore mountains and several locations in the Lower Mainland (University of British Columbia Research Forest at Haney, Chilliwack River valley). PACIFIC WATER SHREW has dark-brown ventral fur, and its skull has a rostrum down-curved in side view.

NATURAL HISTORY

WESTERN WATER SHREWS are found in habitats ranging from low-elevation forest and open wetlands near sea level to high alpine. It has been recorded from elevations as high as 2,318 metres at Mount Assiniboine Provincial Park in the Rocky Mountains and 2,380 metres on Old Glory Mountain near Rossland in the Monashee Mountains. Strongly associated with wet habitats, WESTERN

Capture site at Veitch Creek on Vancouver Island.

WATER SHREWS seem to be most abundant near fast-flowing mountain streams and streams below AMERICAN BEAVER dams where there are many rapids and small riffles. It is frequently found in areas where there are rocks and boulders in and alongside streams, and tree roots in overhanging ledges. In these habitats, it is rarely found more than a few metres from the edge of the stream. Other wetland habitats used are wet meadows and alder-willow thickets bordering ponds and lakes. Known occurrences of the Vancouver Island population are from elevations below 500 metres, with the exception of an observation at 1,000 metres elevation (Bedwell Lake at Strathcona Provincial Park). The lack of records from higher alpine habitats on the island is probably the result of sampling bias.

There are no estimates for population numbers. No information was given on his trapping effort, but Kenneth Racey reported this shrew as being common in the grass-sedge meadows bordering Chilcotin Lake capturing 16 in the summer of 1931. But in most small-mammal trapping studies, WESTERN WATER SHREW represented a small proportion of the total captures, and it appears to be one of the rarer shrews, even in its ideal habitat. In 1992, Gustavo Zuleta and Carlos Galindo-Leal sampled 55 locations in the Lower Mainland with more than 19,000 trap nights. They captured 999 small mammals but only one WESTERN WATER SHREW at Lawson Creek at 430 metres elevation in the North Shore mountains. In BC, the dominant shrews in communities with WESTERN WATER SHREWS are DUSKY SHREW, VAGRANT SHREW or CINEREUS SHREW.

Much of what we know about this animal's behaviour is from Sorenson's study of 13 captive WESTERN WATER SHREWS. Although active at all hours, his animals had two peaks of activity, one just before dawn and the other just after dusk. It readily dived into water and remained submerged for short periods, from a few to 20 seconds; captive shrews can tolerate forced dives of up to 48 seconds. Propulsion in water is by paddling with the feet. Air bubbles trapped in the fur create an important layer of insulation and give the shrew a silvery appearance underwater. Air bubbles trapped in the stiff hairs of the hind feet provide some buoyancy enabling this species to walk briefly on the surface of water. Immediately after a swim, the shrew shakes off the water and vigorously grooms its fur with its hind feet.

Captive WESTERN WATER SHREWS constructed nests from dried vegetation and shredded cloth. They made a depression with their feet and legs, and formed the walls of the nest with their muzzles. Most nests were about eight centimetres in diameter and were located under logs or in cavities of hollow logs. Tunnels constructed by other small mammals were used, but they were capable of digging their own tunnels with their front feet. In captivity WESTERN WATER SHREWS are aggressive, and fighting is common. When two meet, they emit squeaking sounds and often rise up on their hind legs, displaying their light-coloured bellies. Fights are short but intense. The combatants slash each other with their teeth and usually clinch in a tight ball. They often injure their heads and tails in these fights.

WESTERN WATER SHREW's senses are poorly studied, but they presumably use their vibrissae to detect water movements created by their prey as well as using underwater sniffing similar to the AMERICAN WATER SHREW (page 151). Out of water, captive WESTERN WATER SHREWS were able to hear sounds up to three metres away. While moving or exploring their cage, they continually emitted rapid squeaks. High-frequency sounds (25 to 60 kilohertz) could be used for echolocation, but this has not been confirmed. The sense of smell is well developed for detecting odours out of water.

No data are available on the food habits for BC populations. The most detailed information for animals in the wild is from a study of WESTERN WATER SHREWS living in fast-flowing streams in the Rocky Mountains of Montana. There, aquatic insects particularly caddisfly larvae, stonefly nymphs and mayfly nymphs formed the bulk of the diet. Other invertebrates eaten included crane fly and midge larvae, spiders and earthworms. The soft food items are consistent with the moderate bite force of this shrew. There are several reports of it eating small fish such as sculpins, salmon parr and juvenile fish in provincial fish hatcheries. Other observations from the Pacific Northwest include a WESTERN WATER SHREW capturing a PACIFIC GIANT SALAMANDER larva eight centimetres long and a COASTAL TAILED FROG. Captive animals had brief but frequent feeding periods, lasting only 30 to 90 seconds, and in intervals of about 10 minutes;

WESTERN WATER SHREW can survive up to three hours without food. Captive animals captured small fish less than 15 centimetres in length that they killed by bites directly behind the head. Other prey eaten in captivity included snails and dead rodents. Fish, suet, oatmeal and insects were hoarded in a hollow log or buried in the cage for later consumption. The shrews consumed or stored an average of 14.6 grams of food per day.

In Montana, the breeding season extends from December to September. Male WESTERN WATER SHREWS appear to begin breeding earlier than males of other shrew species. Males with enlarged testes were observed as early as December, and the earliest date for pregnant females was February. Males do not breed in their first summer, and young-of-the-year males probably reach sexual maturity in December. Older males (born the previous year) are not sexually active until February. Successful pregnancy is rare in young-of-the-year females, and most pregnant females are adults born the previous year. Females produce two or three litters, with five or six embryos most common. The few observations for BC consist of three pregnant females captured May 28, August 18 and September 11. The maximum life span for this species is about 18 months.

Major predators include the introduced AMERICAN BULLFROG, owls and fish. Predation by an AMERICAN MARTEN and a BELTED KINGFISHER were observed in the Yukon. Similar to the PACIFIC WATER SHREW, this species is prone to drowning in minnow traps used in fish surveys.

CONSERVATION STATUS

Although considered secure in mainland BC, the Vancouver Island subspecies is on the provincial Blue List because of its rarity and potential threats of habitat loss from the harvesting of riparian forest and urban development. However, inadequate data on population numbers and trends, distribution, habitat affinities and taxonomic status prohibit a rigorous conservation status assessment of the Vancouver Island race. Because it is the only known coastal island population of this species in BC, we recommend that the Vancouver Island subspecies warrants treatment as a distinct conservation unit until its taxonomy is resolved.

SYSTEMATICS AND TAXONOMY

In the 1996 edition of this handbook, the water shrew was treated as a single species *Sorex palustris* found across North America. Preliminary genetic studies applying mitochondrial DNA revealed cordilleran, boreal and eastern genetic lineages. A subsequent study with broader geographic sampling and applying both mitochondrial and nuclear gene sequences concluded that the three lineages are sufficiently distinct to warrant recognition as distinct species: EASTERN WATER SHREW, WESTERN WATER SHREW and AMERICAN WATER SHREW. The two latter species are found in BC. The closest relative of the WESTERN

WATER SHREW is the PACIFIC WATER SHREW. Two subspecies are recognized; both occur in the province.

- *Sorex navigator brooksi* Anderson. An insular race restricted to Vancouver Island; the type locality is Black Creek. Rudolph Anderson described this subspecies from a single specimen on the basis of its dark colour especially the dusky-brown underparts. The Vancouver Island population is smaller in body measurements, but our research showed no clear differences from mainland WESTERN WATER SHREWS in cranial/dental measurements. DNA sequences revealed minor divergence from mainland populations, but curiously, two subgroups of Vancouver Island water shrews were evident in the phylogenetic tree, raising the possibility of multiple colonizations from the mainland in the past. A comprehensive genetic study comparing adjacent coastal mainland populations with the Vancouver Island population is required to determine the taxonomic validity of this subspecies. Measurements: total length 137 mm (123–154 mm), $n=35$; tail vertebrae 68 mm (59–76 mm), $n=34$; hind foot 19 mm (18–21 mm), $n=35$; mass 10.3 g (6.1–17.4 g), $n=33$.
- *Sorex navigator navigator* (Baird). Occurs across the entire species's range except for Vancouver Island. Populations in Arizona show some differences in DNA and may warrant recognition as a separate subspecies. Measurements: total length 152 mm (133–180 mm), $n=214$; tail vertebrae 75 mm (62–88 mm), $n=214$; hind foot 20 mm (16–22 mm), $n=214$; mass 10.9 g (7.5–21.5 g), $n=55$.

REMARKS

With the recent taxonomic revision of water shrews, some published literature applies to the AMERICAN WATER SHREW or EASTERN WATER SHREW. For this account, we only used studies done with water shrews from the known range extent of WESTERN WATER SHREW.

A small sculpin fish and a PACIFIC GIANT SALAMANDER larva captured by a WESTERN WATER SHREW in Washington State appeared to be paralyzed, raising the possibility that the saliva of this shrew is toxic (see page 25).

The species name *navigator* means "skilled or engaged in navigation." The *brooksi* subspecies name is an eponym for Allan Brooks, a BC naturalist who collected the type specimen. Evidently, Brooks found the specimen dead on the highway bridge at Black Creek and submitted it to the National Museum of Canada (now the Canadian Museum of Nature) in Ottawa, resulting in its description and naming as a new subspecies.

REFERENCES

Anderson (1934); Beneski and Stinson (1987); Calder (1969); Conaway (1952); Craig (2004); Hope et al. (2014); Jung (2016); Mycroft, Shafer and Stewart (2011); Nagorsen, Panter and Hope (2017); Nussbaum and Maser (1969); O'Neill, Nagorsen and Baker (2005); Racey (1936); Sorenson (1962); Svihla (1934); Zuleta and Galindo-Leal (1994).

Dusky Shrew

Sorex obscurus

Other common name: Navigator Shrew

DESCRIPTION

DUSKY SHREW is a medium-sized shrew with brown dorsal fur shading gradually to brown-to-grey ventral fur in summer pelage. Some individuals are faintly tricoloured. The winter pelage is darker with brown-mixed-with-grey dorsal fur and smoky-grey ventral fur. The tail is bicoloured, but not distinctly. The second to fifth digits on the hind feet have more than four pairs of toe pads. The moult to summer pelage usually begins in early May, with most individuals in full summer pelage by the first week of June; however, in some individuals the moult is delayed until late June or July. The acquisition of winter fur begins in September and is usually complete by mid-October.

Cranial/dental traits: 32 teeth; incisors 1/1, unicuspids 5/1, premolars 1/1, molars 3/3; upper incisor with a medium-sized medial tine located well within the pigmented area; third upper unicuspid shorter than fourth; upper unicuspids with a strongly pigmented ridge which reaches the cingulum; lower incisor has three denticles with continuous pigment on the labial side covering all three denticles, with lingual pigment reaching the second or third denticle; tiny to medium-sized postmandibular foramen present in one or both sides of the dentaries in about 60 per cent of the individuals we examined.

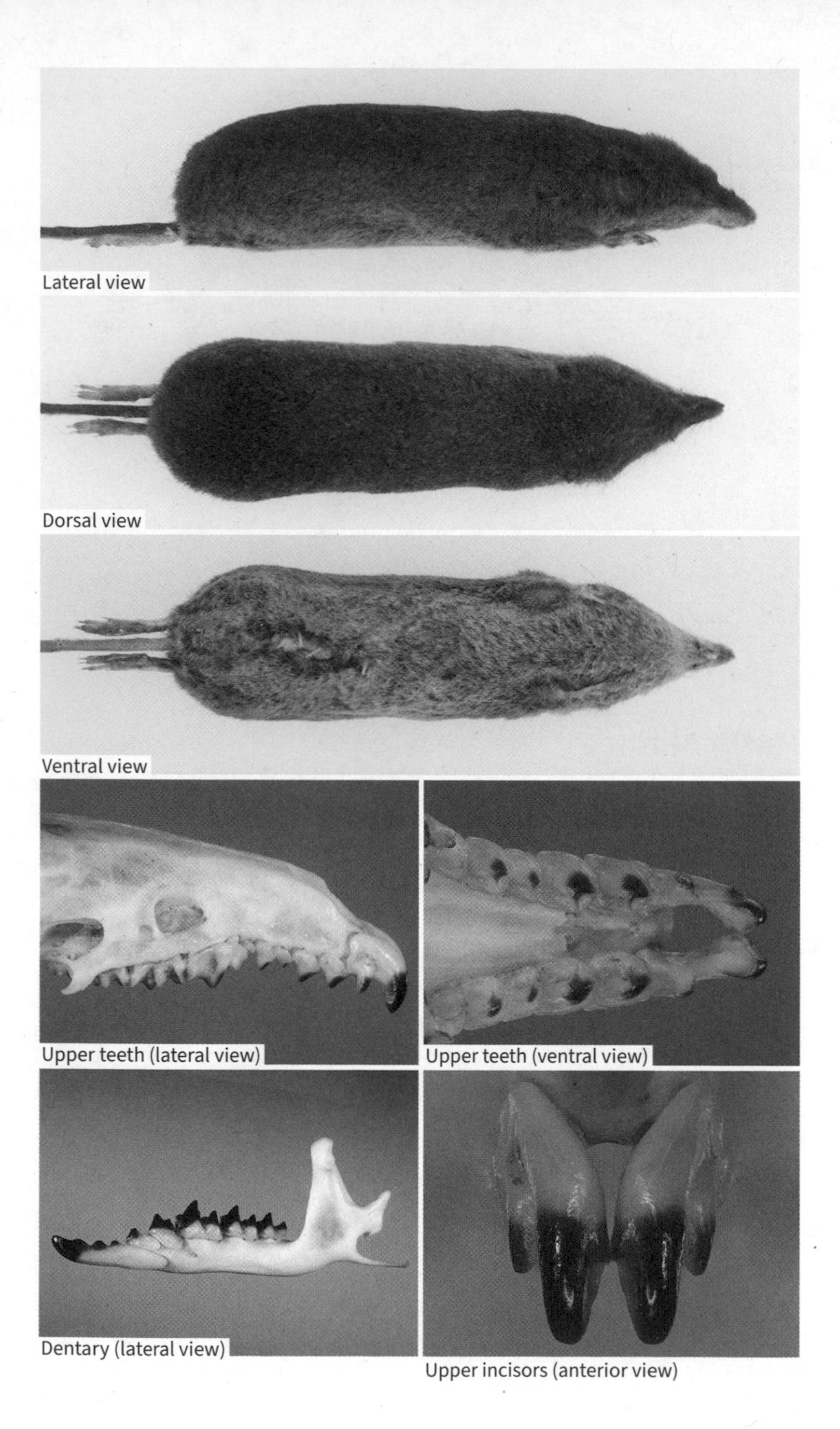
Lateral view
Dorsal view
Ventral view
Upper teeth (lateral view)
Upper teeth (ventral view)
Dentary (lateral view)
Upper incisors (anterior view)

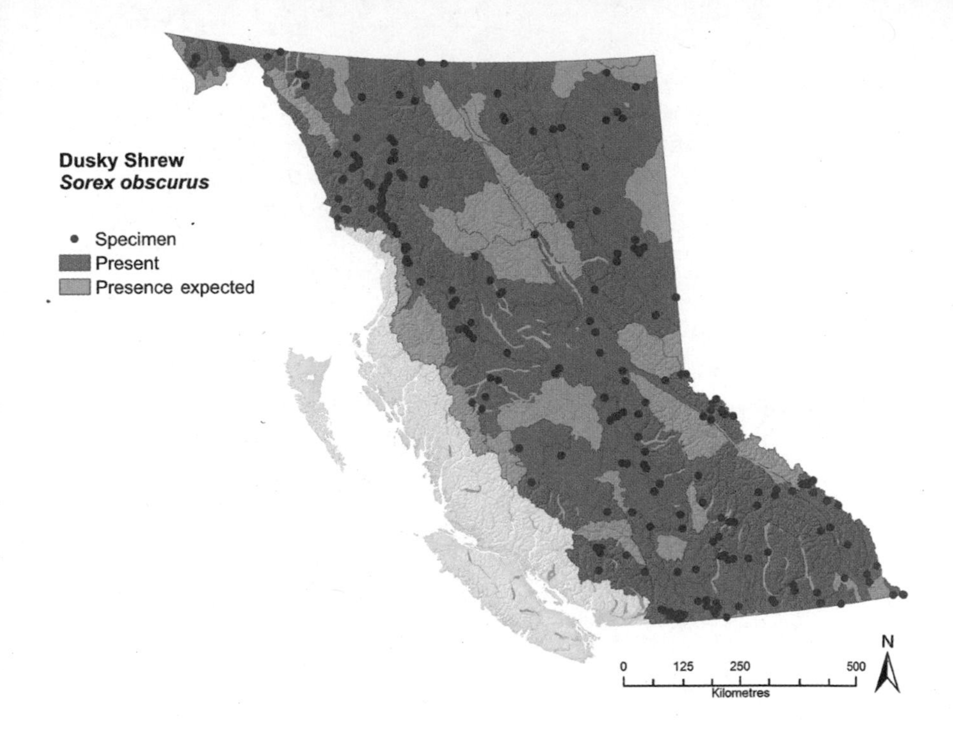

DISTRIBUTION

DUSKY SHREW ranges from Alaska and western Canada to as far south as Idaho, Montana and eastern Oregon. In BC, it occupies the entire mainland east of the Coast and Cascade Mountains. Westernmost occurrences of individuals identified from DNA are at Iskut and Dease Lake in northern BC. Specimens from Alpha Lake near Whistler, Bennett and the Great Glacier on the Stikine River near the Alaskan border were identified from skull morphology as the *obscurus* subspecies by Lois Alexander, suggesting that the range extends well into some valleys of the Coast Mountains. More genetic sampling is essential to delimit the western limits of its range and to determine the extent that its distribution overlaps with that of PACIFIC SHREW.

MEASUREMENTS

	Mean	Range	Sample size
Total length:	110 mm	90–129 mm	*n* = 376
Tail vertebrae:	47 mm	37–59 mm	*n* = 374
Hind foot:	13 mm	10–16 mm	*n* = 373
Mass—females:	5.5 g	3.9–10.0 g	*n* = 37
Mass—males:	6.5 g	4.0–9.0 g	*n* = 39

MORPHOLOGICALLY SIMILAR SPECIES

Most difficult to distinguish from DUSKY SHREW is the closely related PACIFIC SHREW. PACIFIC SHREW generally has a darker pelage, but both share similar external and cranial/dental traits with some overlap in their measurements. Positive identification requires a genetic sample. Other shrews likely to be misidentified as DUSKY SHREWS are ARCTIC SHREW, CINEREUS SHREW, TUNDRA SHREW and VAGRANT SHREW. ARCTIC SHREW and TUNDRA SHREW have a prominent tricoloured pelage, a third upper incisor that is larger than the fourth, and an upper first molar with a pigmented hypocone. CINEREUS SHREW is smaller in size (skull length less than 16.0 millimetres), and its third upper unicuspid is equal or larger than the fourth. With a third upper unicuspid that is smaller than the fourth similar to DUSKY SHREW, VAGRANT SHREW is challenging to discriminate. DUSKY SHREW averages larger in skull size and tail length, but the two species overlap substantially in most measurements. VAGRANT SHREW has no more than four pairs of toe pads on the second to fifth digits of its hind foot, and the medial tine on its upper incisor is small, positioned above the pigmented region and separated from the main pigmented area by a pale border.

NATURAL HISTORY

DUSKY SHREW occupies a wide spectrum of habitats from valley bottoms to alpine: forests, steppe grassland, grassy meadows, riparian areas, shrub thickets, and subalpine and alpine meadows. There are captures at 2,240 metres in the Purcell Mountains (Paradise Mine) and 2,286 metres in the Rocky Mountains (Mount Assiniboine Provincial Park). It is also common in disturbed areas such as recent burns and logged forests. Several studies have reported captures in vole runways. Two shrews commonly associated with the DUSKY SHREW in BC are VAGRANT SHREW and CINEREUS SHREW. In high-elevation habitats, the DUSKY SHREW is often the dominant shrew species, but this is not observed in forested habitats. In the dry DOUGLAS-FIR forest habitats at Opax Mountain near Kamloops, capture rates were higher for CINEREUS SHREW.

CONSERVATION STATUS

With its broad habitat associations and large range extent, DUSKY SHREW is considered secure in BC.

SYSTEMATICS AND TAXONOMY

Molecular genetic studies based on mitochondrial DNA demonstrated that the DUSKY SHREW as described in the 1996 edition of this handbook consists of three genetic clades: northern continental, southern continental, and coastal. Neal Woodman formally recognized the three clades as distinct species, with the northern continental clade classified as DUSKY SHREW (*Sorex obscurus*), southern continental as MONTANE SHREW (*Sorex monticola*) and the coastal clade as PACIFIC SHREW (*Sorex pacificus*). With some reservations, we follow his taxonomy here recognizing both the DUSKY SHREW and PACIFIC SHREW in the province. Nonetheless, the genetic study by John Demboski and Joe Cook was based on few samples from widely scattered locations. Broader geographic sampling applying both nuclear and mitochondrial DNA markers is critical to determine the degree of genetic divergence among the clades and if they warrant treatment as full species.

Four subspecies are recognized for DUSKY SHREW with one found in BC. Although *Sorex obscurus alascensis* could be found in the extreme northwestern corner of BC at the Alaskan border, specimens from the Haines Highway and Tats Lake west of the Tatshenshini River were assigned to the subspecies *Sorex obscurus obscurus* by Lois Alexander.

- *Sorex obscurus obscurus* Merriam. A widespread race ranging from Alaska throughout the cordillera of western Canada and south to Idaho and Montana in the United States. In BC, it is found across the entire mainland east of the Coast Mountains and Cascade Mountains.

REMARKS

The species name *obscurus* from Latin means "obscure" or "little known." This is an appropriate name given this species's changeable taxonomy and recognition under three different species names. It was named and described as *Sorex obscurus* in 1895. But subsequently, it was treated as a subspecies of both *Sorex vagrans* and *Sorex monticolus* (the spelling was changed to *monticola* in later publications) until the recent revision by Neal Woodman restored the species name *Sorex obscurus*.

With the taxonomic split of the former Dusky Shrew, most published studies on population size, movements, diet and reproduction apply to PACIFIC SHREW. The DUSKY SHREW is another candidate for a focused research study on its general biology in BC.

REFERENCES

Alexander (1996); Demboski and Cook (2001); Smith and Belk (1996); Van Tighem and Gyug (1984); Woodman (2018).

Pacific Shrew *Sorex pacificus*

DESCRIPTION

PACIFIC SHREW is a medium-sized shrew with dorsal pelage that varies from dark brown to nearly black in some coastal populations, and shading gradually to brown or grey ventral fur. The tail is bicoloured, but not distinctly. Winter pelage tends be more grey with longer fur. The moult to winter pelage begins in late September. Females acquire their summer pelage in late March before pregnancy; males retain their winter pelage until May. Prominent flank glands develop in breeding males in January or February; they are not evident externally in immature males or females. The second to fifth digits on the hind feet have more than four pairs of toe pads.

Cranial/dental traits: 32 teeth; incisors 1/1, unicuspids 5/1, premolars 1/1, molars 3/3; medial edge of upper incisors often straight in front view; upper incisor with medium-sized medial tine positioned from upper edge to within the pigmented area; third upper unicuspid shorter than fourth; upper unicuspids with a weak to strongly pigmented ridge that extends to the cingulum; lower incisor has three denticles with continuous pigment on the labial side covering all three denticles, with the lingual pigment usually extending to the second or third denticle, and extending to the first in some weakly pigmented specimens. A postmandibular foramen is present in about one-quarter of individuals we examined; if present, it is tiny to medium-sized and usually on one dentary. There is geographic variation in both the intensity of pigmentation and the size of the medial tines of the upper incisor, with some coastal populations from central and northern BC having faint pigmentation on the face of the upper incisor with small medial tines.

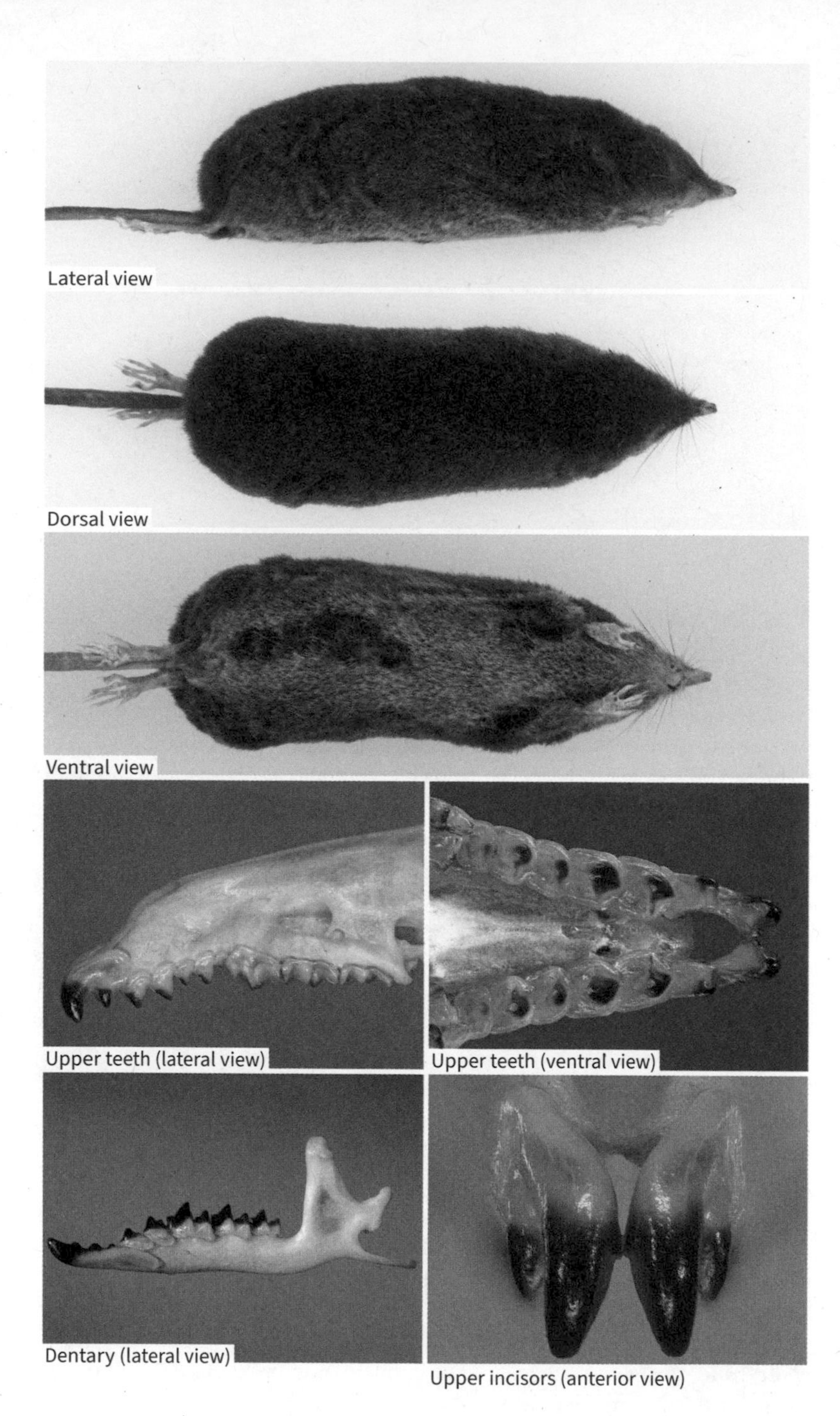

Upper incisors (anterior view)

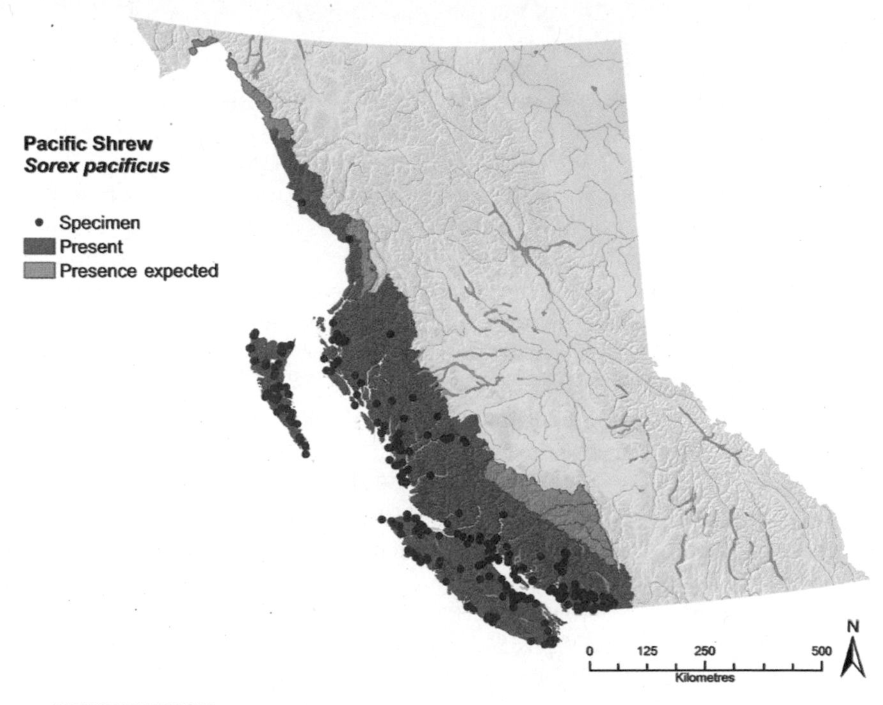

DISTRIBUTION

PACIFIC SHREW ranges from the Cascade Mountains and coastal regions of Oregon through western Washington, western BC and southeastern Alaska. In BC, it inhabits the entire coastal mainland and about 100 islands including Vancouver Island, the islands of the Great Bear Rainforest and the islands of Haida Gwaii. Although widespread on Vancouver Island and present on the northern Gulf Islands and islands off the west coast of Vancouver Island, it is absent from the southern Gulf Islands. The smallest islands known to support populations of PACIFIC SHREW are about one to four hectares in area. Easternmost occurrences in BC of PACIFIC SHREW identified from DNA are Maple Ridge in the lower Fraser River valley, Lakelse Lake south of Terrace, Nass River and the Great Glacier on the Stikine River near the Alaskan border. Until more genetic sampling is done, the eastern limits of its range in the Coast and Cascade Mountains is unknown.

MEASUREMENTS

	Mean	Range	Sample size
Total length:	122 mm	93–145 mm	$n = 1501$
Tail vertebrae:	54 mm	35–70 mm	$n = 1502$
Hind foot:	14 mm	10–19 mm	$n = 1504$
Mass—females:	6.9 g	3.9–13.0 g	$n = 163$
Mass—males:	6.9 g	3.8–11.3 g	$n = 157$

MORPHOLOGICALLY SIMILAR SPECIES

The most difficult species to distinguish from PACIFIC SHREW is the closely related DUSKY SHREW. See its species account (page 137) for identification issues and areas where the two species may co-occur. Other shrews likely to be misidentified as PACIFIC SHREWS are CINEREUS SHREW, TROWBRIDGE'S SHREW and VAGRANT SHREW. CINEREUS SHREW is smaller and has a third upper unicuspid equal or larger than the fourth. TROWBRIDGE'S SHREW and VAGRANT SHREW have a third upper unicuspid smaller than the fourth similar to PACIFIC SHREW. TROWBRIDGE'S SHREW is identified by its distinctly bicoloured tail with a white ventral surface that contrasts with the dark-brown belly fur, the pigmented ridge on its upper unicuspid teeth separated from the cingulum by a longitudinal groove, and the presence of a medium to large postmandibular foramen in both dentaries. Discriminating PACIFIC SHREW from VAGRANT SHREW is challenging, especially in old animals with worn teeth or in some coastal populations where the teeth are weakly pigmented. PACIFIC SHREW averages larger in skull size and tail length, but the two species overlap substantially in most measurements. VAGRANT SHREW has no more than four pairs of toe pads on the second to fifth digits of its hind foot, and the medial tine on its upper incisor is small, positioned at or above the pigmented region with the tine bordered by a pale band where it contacts the main pigmented area.

NATURAL HISTORY

PACIFIC SHREW is found in habitats from alpine tundra to coastal forests. Few elevational data exist, but a historical museum specimen from Overlord Mountain in Garibaldi Provincial Park was captured at 1,890 metres. At low elevations in southwestern British Columbia, it is primarily a forest species and is rare in open fields, wet meadows and grassland habitats. It is also common in disturbed areas such as recent burns and logged forests. In coastal WESTERN HEMLOCK forests, PACIFIC SHREW is evenly distributed in all stands from recent clearcuts to old-growth. Forests with a closed canopy, dry soil and abundant woody debris provide optimum habitat. There are few data on habitat use on coastal islands. Charlie Guiguet found that on the isolated Goose Island off the central BC coast, most captures of PACIFIC SHREWS were in meadows and beach debris with few captures in forested habitats or bogs on the island. There is some evidence that on islands of Haida Gwaii, this shrew is more abundant in marine shoreline habitats than inland forests. The preference for marine shoreline habitats on smaller islands may reflect the abundant invertebrate resources in these habitats.

Two shrews commonly associated with PACIFIC SHREW on the BC mainland coast are VAGRANT SHREW and CINEREUS SHREW. In most forested habitats, PACIFIC SHREW is the dominant species in abundance. In the southern coastal region, this shrew avoids open grassland habitats preferred by VAGRANT SHREW,

Shoreline habitat at Haida Gwaii.

but it appears to have a competitive advantage over VAGRANT SHREWS in WESTERN HEMLOCK forests with well-drained acidic soils. CINEREUS SHREW is rarely a dominant shrew in mainland coastal habitats. On Vancouver Island, PACIFIC SHREW coexists with VAGRANT SHREW and WESTERN WATER SHREW, and on a few of the north coast islands, it co-occurs with CINEREUS SHREW. However, on most British Columbian islands with shrew populations, PACIFIC SHREW is the only shrew present.

In her forest plot near Maple Ridge, Myrnal Hawes estimated peak populations at 5 to 12 PACIFIC SHREWS per hectare. The average home range was 1,227 square metres for non-breeding animals and 4,020 square metres for breeding animals. During the breeding season, males occupied larger areas than females. In late summer, discrete territories are established, and the daily movements of neighbouring animals show no overlap.

Remarkably, there are few studies of PACIFIC SHREW's diet. The stomachs of a small sample taken from a population in the Cascade Mountains of Washington contained mostly unidentified invertebrates and conifer seeds; fungi and lichens were found in a few individuals. Hawes speculated that PACIFIC SHREW has larger, more robust teeth than VAGRANT SHREW because it has adapted to feeding on small, hard-bodied invertebrates that would be expected to live in the dry acidic soils that characterize its habitats.

In Hawes's study area, males became sexually active in mid-February and females in early April just after they completed their moult. The first young appeared in mid-May. Females may produce two litters in the breeding season. Litter size ranges from two to seven, with four most common. Males and females do not breed in the summer of their birth. The life span is about 16 to 17 months. More than half of the young die within the first five months of life, and only about 4 per cent will survive from their summer of birth to the following summer. Major predators are owls and mammalian predators. In BC, PACIFIC SHREW remains have been recovered from the digestive tracts or scats of PACIFIC MARTEN and ERMINE.

CONSERVATION STATUS

The conservation status of PACIFIC SHREW has not been assessed by the province. Given its wide distribution and habitat use on the BC coast, the species is not of conservation concern. Nonetheless, four of the seven subspecies listed for the province are races endemic to small BC islands. They should be a high priority for a conservation assessment.

SYSTEMATICS AND TAXONOMY

The splitting of the former Dusky Shrew into three species, with the coastal clade treated as PACIFIC SHREW, is discussed in the DUSKY SHREW account. Lois Alexander recognized 11 subspecies of PACIFIC SHREW; seven are found in BC, with five confined to coastal islands. They were described solely on morphological traits. Only seven specimens from coastal BC were included in the 2001 mitochondrial DNA study of DUSKY SHREW phylogeny. A molecular genetic study applying mitochondrial and nuclear DNA is required to evaluate the validity of these subspecies.

- *Sorex pacificus calvertensis* Cowan. A weakly defined, insular subspecies confined to two widely separated islands on the north coast: Banks and Calvert. The type locality is Safety Cove, on Calvert Island. This race is smaller than *Sorex pacificus longicauda* in skull and body size but similar to *Sorex pacificus elassodon* in size. Measurements: total length 120 mm (106–135 mm), $n=37$; tail vertebrae 53 mm (44–59 mm), $n=37$; hind foot 14 mm (13–15 mm), $n=37$; mass 7.6 g (5.0–9.0 g), $n=18$.
- *Sorex pacificus elassodon* Osgood. An insular race associated with islands of the Alexander Archipelago in southeastern Alaska, Haida Gwaii islands (excluding Kunghit Island), and Porcher and Dewdney Islands. It is smaller than *Sorex pacificus prevostensis* from Kunghit Island, in Haida Gwaii. The type locality is Cumshewa Inlet, on Moresby Island in Haida Gwaii. Measurements: total length 121 mm (95–141 mm), $n=215$; tail vertebrae 53 mm (39–65 mm), $n=214$; hind foot 14 mm (11–18 mm), $n=216$; mass 7.3 g (4.0–13.0 g), $n=80$.
- *Sorex pacificus insularis* Cowan. An insular race confined to Reginald, Athlone (formerly Smyth) and Townsend (see note on page 111) Islands

in the Bardswell Group. The type locality is Smythe Island. Compared to *Sorex pacificus longicauda*, this race is larger in some skull measurements but smaller in others. Measurements: total length 123 mm (111–138 mm), *n* = 61; tail vertebrae 53 mm (46–60 mm), *n* = 61; hind foot 15 mm (13–19 mm), *n* = 61; no data available for body mass.

- *Sorex pacificus isolatus* Jackson. An insular race found on Vancouver Island, Denman Island, Cox Island in the Scott Islands, and many islands off the west coast of Vancouver Island. It is smaller than *Sorex pacificus setosus*, the subspecies found on the southern mainland and islands adjacent to the mainland. The type locality is the mouth of Millstone Creek (that is, Millstone River) in Nanaimo. Measurements: total length 116 mm (93–133 mm), *n* = 186; tail vertebrae 50 mm (36–59 mm), *n* = 186; hind foot 13 mm (10–19 mm), *n* = 185; mass 6.0 g (4.0–9.0 g), *n* = 96.
- *Sorex pacificus longicauda* Merriam. Occurs from Taku, Alaska to Rivers Inlet on the coastal mainland. Eastern limits of the range are on the eastern slopes of the Coast Mountains. It also occurs on many islands: Campania, Campbell, Chatfield, Dufferin, Horsfall, Hunter, West Kinahan, Pitt, Princess Royal, Spider, Swindle and Yeo. Mainland populations are larger in skull and body size than *Sorex pacificus setosus*, the coastal subspecies to the south. Measurements: total length 128 mm (96–145 mm), *n* = 414; tail vertebrae 58 mm (36–70 mm), *n* = 416; hind foot 15 mm (12–17 mm), *n* = 416; mass 6.9 g (4.6–10.9 g), *n* = 48.
- *Sorex pacificus prevostensis* Osgood. Confined to Kunghit Island (formerly Prevost) in Haida Gwaii. It has the largest body mass and skull size of the seven subspecies in BC. Measurements: total length 127 mm (105–142 mm), *n* = 49; tail vertebrae 54 mm (45–61 mm), *n* = 49; hind foot 15 mm (14–15 mm), *n* = 49; mass 9.0 g (5.2–13.0 g), *n* = 28.
- *Sorex pacificus setosus* Elliot. Occurs from the southern coastal mainland from Rivers Inlet to Washington State. Eastern limits are the western slopes of the Coast and Cascade Mountains. It also occurs on a number of coastal islands including Bowen, Cortes, Goose, Hecate, Marina, Maurelle, Quadra, Sonora, Stuart and Texada. The Texada Island population was originally classified as a distinct race, which is no longer considered valid. Measurements: total length 119 mm (94–137 mm), *n* = 539; tail vertebrae 54 mm (35–64 mm), *n* = 539; hind foot 13 mm (10–19 mm), *n* = 540; mass 6.1 g (3.7–11.0 g), *n* = 69.

REMARKS

Although it is the most ubiquitous and abundant shrew on the BC coast, PACIFIC SHREW has been the subject of surprisingly little research. Its diet has not been described, and information on breeding biology is limited to the southern coastal mainland. The numerous BC islands with their wide range of sizes, degree of isolation and ecological complexity provide opportunities for comparative studies of populations, diet, habitat use, genetics and reproduction among insular PACIFIC SHREWS.

The species name *pacificus* refers to the geography of this shrew's range. The name originally was applied to a species restricted to coastal Oregon.

REFERENCES

Alexander (1996); Burles, Edie and Bartier (2004); Craig (1995); Demboski and Cook (2001); Doyle (1990); Foster (1965); George and Smith (1991); Guiguet (1953); Gunther, Horn and Babb (1983); Hawes (1975, 1977); McCabe and McTaggart Cowan (1945); McTaggart Cowan (1941); Smith and Belk (1996).

American Water Shrew *Sorex palustris*

DESCRIPTION

AMERICAN WATER SHREW is a large shrew with dark-grey or black dorsal fur, although some appear dull brown; its ventral fur is silver grey, sometimes washed with brown. Scattered silvery-white hairs are also visible on some animals. The tail is distinctly bicoloured, dark above and white or grey below. Similar to the WESTERN WATER SHREW, the ventral surface of the tail for about the last centimetre from the tip has a raised fleshy ridge about one millimetre high fringed by stiff hairs. This forms a widening of the tail compared with other shrews and presumably functions to aid propulsion in water. The chin is pale white or grey. The hind feet are large with a fringe of stiff hairs about one millimetre long on the outer margins. Smaller stiff hairs occur on the front feet. Males in reproductive condition have prominent flank glands.

Cranial/dental traits: 32 teeth; incisors 1/1, unicuspids 5/1, premolars 1/1, molars 3/3; upper incisor with medial tine that is medium to large in size, distinct and included in the anterior pigment; third upper unicuspid distinctly shorter than fourth; upper unicuspids with a weak to strongly pigmented ridge that extends to the cingulum in about 50 per cent of individuals or lacking a pigmented ridge in the other 50 per cent; lower incisor has three denticles with continuous pigment on the labial side covering two or three denticles, with the pigmented area on the lingual side extending to the second or third denticle. A tiny to medium-sized postmandibular foramen was present usually in one dentary for about half the individuals we examined.

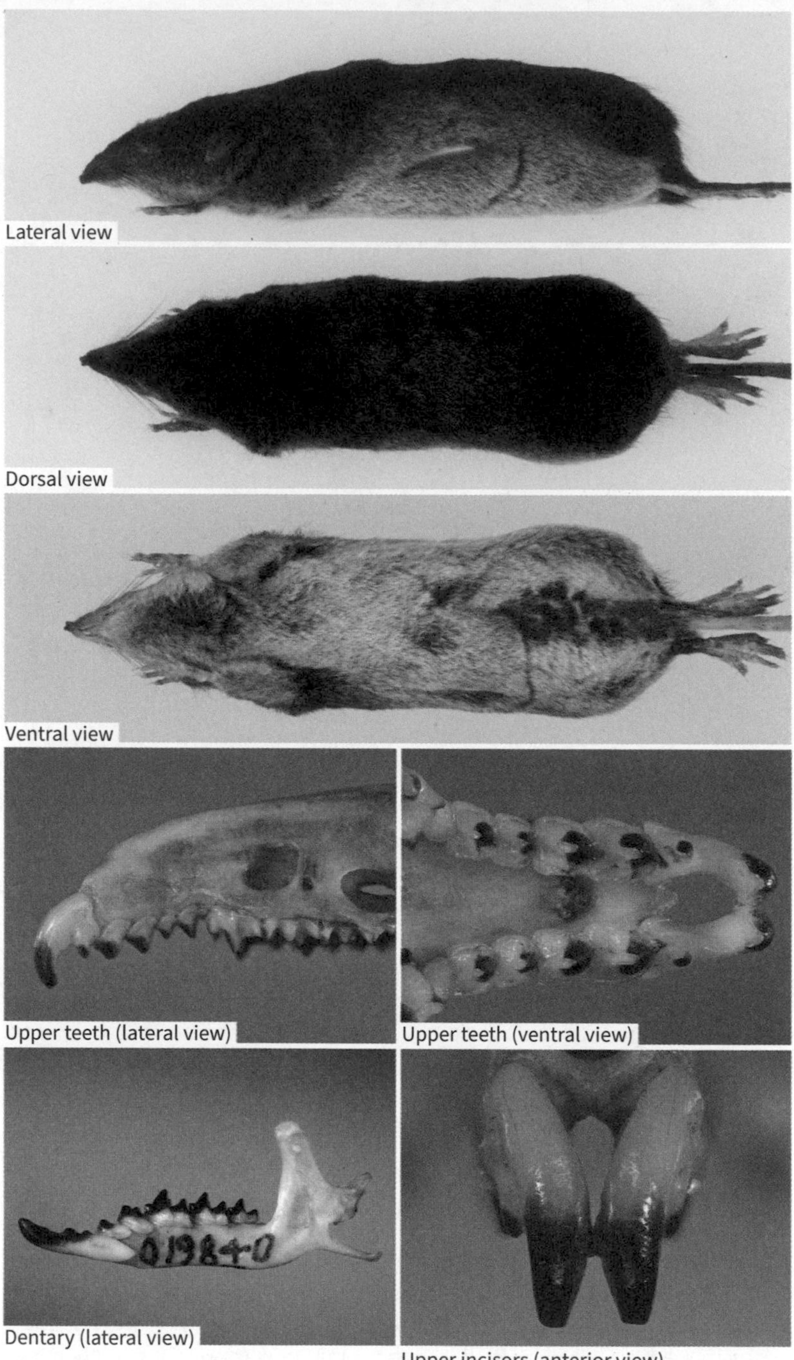
Lateral view
Dorsal view
Ventral view
Upper teeth (lateral view)
Upper teeth (ventral view)
Dentary (lateral view)
Upper incisors (anterior view)

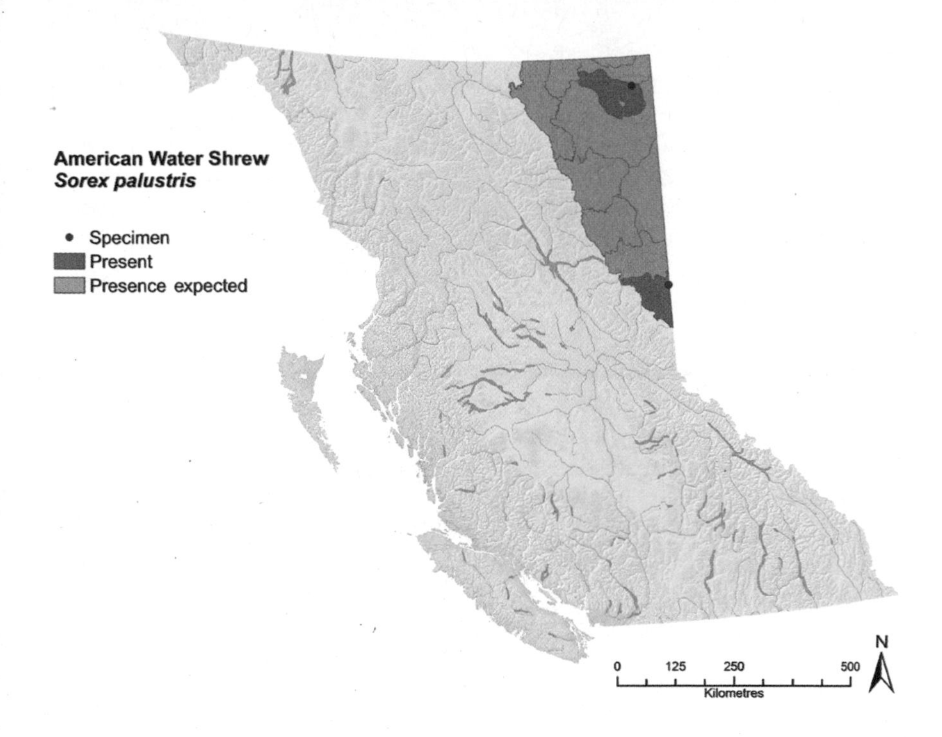

DISTRIBUTION

AMERICAN WATER SHREW ranges from northwestern Ontario to northern Alberta, Northwest Territories and northeastern BC. In BC, it is known by only two historical museum specimens in the collections of the Royal BC Museum that were captured at Tupper Creek and Kotcho Lake. Western limits of the BC range are unknown, but it likely reaches the foothills of the northern Rocky Mountains.

MEASUREMENTS

Measurements are based on 2 BC specimens and 40 specimens from central and northern Alberta.

	Mean	Range	Sample size
Total length:	146 mm	128–172 mm	$n = 41$
Tail vertebrae:	66 mm	56–76 mm	$n = 41$
Hind foot:	19 mm	17–21 mm	$n = 42$
Mass—females:	12.5 g	9.6–15.1 g	$n = 12$
Mass—males:	13.4 g	8.9–17.3 g	$n = 14$

MORPHOLOGICALLY SIMILAR SPECIES

Within its known BC range extent, AMERICAN WATER SHREW is easily distinguished from other shrews by its large size, distinctive pelage colour and stiff hairs on the hind feet. Although their known occurrences are separated by about 200 kilometres, AMERICAN WATER SHREW and WESTERN WATER SHREW may co-occur in the foothills of the northern Rocky Mountains. They cannot be discriminated from external features, and identification will require cranial/dental measurements taken from a skull of a voucher specimen (see Identification Key to Cranial/Dental Traits, page 52) or a genetic sample.

NATURAL HISTORY

The most detailed habitat data are from Manitoba, where AMERICAN WATER SHREW is relatively common. Most captures there were in grass-sedge meadows and alder-willow thickets bordering ponds and lakes. It also inhabits forests with water bodies nearby. Robert Wrigley and colleagues noted a strong association with the presence of this shrew and dams built by AMERICAN BEAVERS. Running water below the dams and marshy meadows surrounding beaver ponds provide ideal habitat. Although rare in bogs, a population in Manitoba occupied a bog in a TAMARACK forest with LABRADOR TEA, DWARF BIRCH and sphagnum moss ground cover situated nearly a kilometre from the nearest lake or stream. Habitat data for BC are scanty. The Kotcho Lake specimen was taken near an old beaver pond; the Tupper Creek capture was in a vole runway in a marshy pond border. Elevational range in BC is unknown.

Home ranges of 0.2 and 0.3 hectares were reported for two individuals living in a TAMARACK bog in Manitoba, but the movements of this species have not been well documented. No estimates exist for population numbers. In most small-mammal trapping studies, AMERICAN WATER SHREW represented a small proportion of the total captures. The dominant shrew in communities with AMERICAN WATER SHREWS in BC is DUSKY SHREW or CINEREUS SHREW.

Until recently, AMERICAN WATER SHREW's senses were not well understood. It was generally assumed that touch, particularly with the sensitive nose and vibrissae, was used to locate aquatic insects and other invertebrates on stream bottoms. There was also speculation that it uses echolocation, sonar or some form of electro-reception to locate its prey. Laboratory studies on captive AMERICAN WATER SHREWS at the University of Manitoba revealed that this shrew finds its prey by both underwater sniffing, where water is taken in and expired through the nostrils while submerged, and detecting water movements of prey with the sensory vibrissae on its snout. Vision seems to play little role in foraging. With these adaptations, AMERICAN WATER SHREW can find prey in water even in total darkness. The laboratory studies found no evidence that it responds to electric fields or uses ultrasonic or audible sounds underwater

Beaver pond habitat.

or above ground while foraging. Mean voluntary dive time for captive animals was about 6 seconds; the longest dive time was 24 seconds.

There are few dietary data, but based on what we know about food habits of other water shrew species, aquatic invertebrates likely form the bulk of the diet. In Minnesota, slugs and earthworms were major prey items with spiders, crickets and lepidopteran larvae minor prey. An AMERICAN WATER SHREW, observed hunting in a small stream pool in Manitoba, sat on a rock watching for COMMON SHINER minnows passing nearby. Three of these fish (about six centimetres in length) were caught using the same technique. The shrew dived into the pool attacking a minnow in its belly area with a series of swift bites, then carried it back to the rock and consumed the head and viscera. In TAMARACK bogs, the bulk of its diet is made up of carabid beetles, moth and butterfly larvae, and the pupae and larvae of sawflies. A physiological study on captive AMERICAN WATER SHREWS found that they must eat the equivalent of 14 to 15 BROOK STICKLEBACK minnows per day to meet their daily energy requirements.

Minnows of SPOTTAIL SHINER are potential prey in northeast BC.

We could find no published information on reproduction of AMERICAN WATER SHREWS in the wild, but presumably its breeding season and age at reproductive maturity are similar to that described for WESTERN WATER SHREW. Two captive females gave birth to litters of three and four young after a gestation period of about 21 days. At birth the young are hairless. The juvenile pelage is fully developed by 18 days, and the fringe of hairs on the feet appears by 16 days. In captivity, the young grow very fast (0.51 grams per day). At 16 days they are able to walk, and they are capable of diving in an aquarium at 25 days. By 40 to 45 days they are proficient divers.

Predators of the AMERICAN WATER SHREW include snakes, large frogs, owls and fish. The maximum life span for this species in the wild is about 18 months, although in captivity, it may live up to 28 months.

CONSERVATION STATUS

AMERICAN WATER SHREW is on the province's Blue List as a species of concern. With only two captures, its distribution, habitat and population trends in BC are essentially unknown. Its range is in a region undergoing significant development and habitat change from oil and gas exploration, flooding from the Site C hydroelectric dam, and forest harvesting.

SYSTEMATICS AND TAXONOMY

See the WESTERN WATER SHREW account for details on the recent taxonomic revision splitting water shrews into three species. The closest relative of AMERICAN WATER SHREW is EASTERN WATER SHREW, found in eastern North America. DNA sequencing suggests they diverged from a common ancestor some 290,000 years ago.

Two subspecies are recognized with one occurring in BC. Although populations from the western Great Lakes region were classified as a subspecies of AMERICAN WATER SHREW (*Sorex palustris hydrobadistes*) in the recent taxonomic revision, more genetic sampling is required to verify if this population aligns with AMERICAN WATER SHREW or EASTERN WATER SHREW.

- *Sorex palustris palustris* Richardson. A broad range from northern Ontario to northern Alberta, Northwest Territories and northeastern BC. We found that AMERICAN WATER SHREWS from Alberta had larger skulls than those from Manitoba, suggesting a possible west-to-east trend of decreasing size in this subspecies.

REMARKS

AMERICAN WATER SHREW is likely widely distributed across northeastern BC. Only two confirmed occurrences in the province reflect sampling bias rather than rarity. Both captures were taken opportunistically during general vertebrate surveys by the Royal BC Museum. Several expeditions by other museums sampled small mammals along the BC portion of the Alaska Highway but failed to capture a single water shrew. No details are available on sampling effort or habitats sampled during these surveys. Recent small-mammal inventories in the Peace River region including surveys for the BC Hydro Site C dam assessment focused primarily on forested habitats, with none in wetlands or riparian habitats specifically targeting water shrews. AMERICAN WATER SHREW is another candidate for a focused research study in BC.

No genetic samples are available for the two BC historical museum specimens, but we found that their cranial/dental measurements unequivocally identify them as AMERICAN WATER SHREWS rather than WESTERN WATER SHREWS.

The species name *palustris* is from Latin meaning "swampy" or "marshy."

REFERENCES

Beneski and Stinson (1987); Buckner (1970); Buckner and Ray (1968); Catania (2013); Catania, Hare and Campbell (2008); Gusztak (2008); Gusztak and Campbell (2004); Gusztak, MacArthur and Campbell (2005); Hope et al. (2014); McTaggart Cowan (1939); Mycroft, Shafer and Stewart (2011); Nagorsen, Panter and Hope (2017); Whitaker and Schmeltz (1973); Wrigley, Dubois and Copland (1979).

Preble's Shrew *Sorex preblei*

DESCRIPTION

Next to WESTERN PYGMY SHREW, PREBLE'S SHREW is our smallest shrew. It has a dorsal pelage that is brown in summer and dark grey in winter. The underside is grey. The tail is lightly bicoloured, brown above with a paler ventral side. The third and fourth digits of the hind foot have five pairs of toe pads.

Cranial/dental traits: 32 teeth; incisors 1/1, unicuspids 5/1, premolars 1/1, molars 3/3; upper incisor with a well-developed medium-sized medial tine positioned well within the pigment edge; first upper unicuspid with weak or interrupted pigment to cingulum; third upper unicuspid equal or taller than the fourth; lower incisor has three denticles with pigmentation on the labial side covering two or three denticles, with lingual pigment extending to the second denticle; dentary lacks a postmandibular foramen.

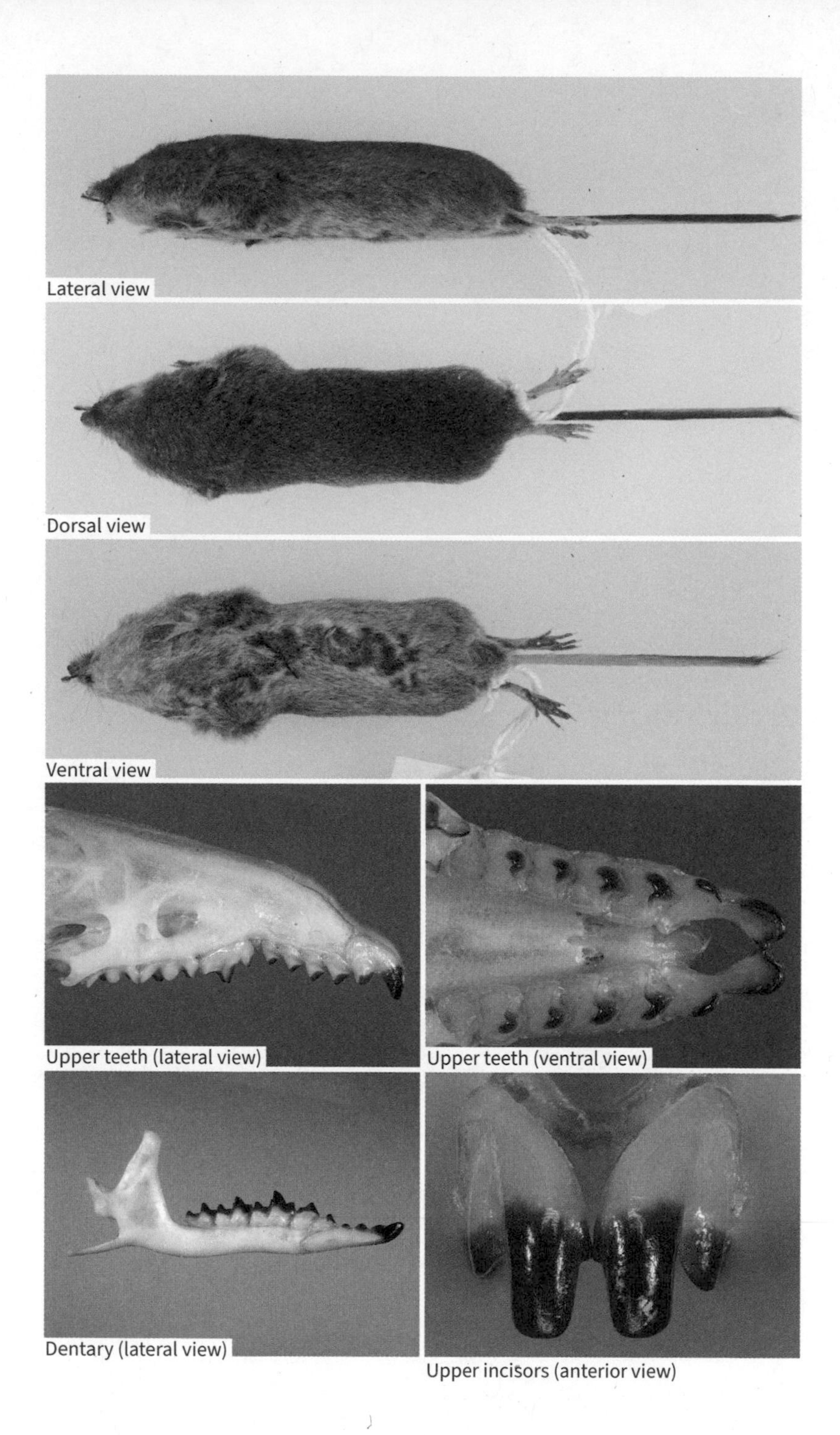

Upper incisors (anterior view)

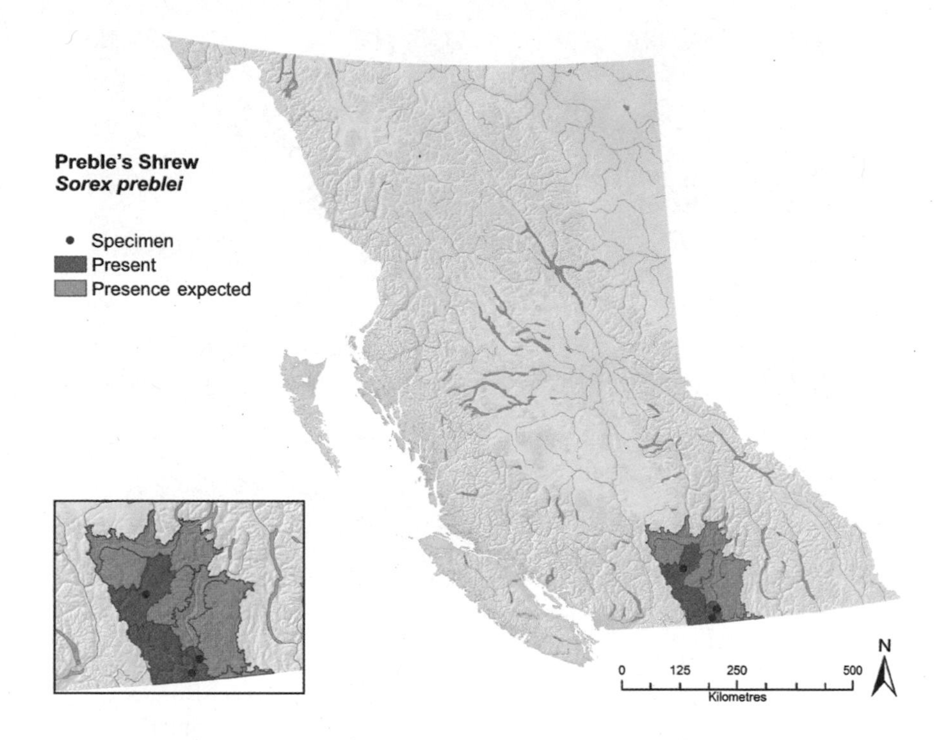

DISTRIBUTION

The distribution of PREBLE'S SHREW consists of widely scattered localized occurrences across western North America from California and New Mexico north to Washington, Montana and British Columbia. Its peculiar spotty range with large gaps is probably a sampling bias. In BC it is known from only four locations. Three are in the south Okanagan: two east of Vaseux Lake and one on Mount Kobau. The other capture site is about 150 kilometres northwest of the Okanagan sites in the Hamilton Commonage, east of Nicola Lake.

MEASUREMENTS

Measurements are based on three BC specimens and nine specimens from Washington State.

	Mean	Range	Sample size
Total length:	86 mm	71–94 mm	$n=12$
Tail vertebrae:	35 mm	32–38 mm	$n=12$
Hind foot:	10 mm	9–11 mm	$n=10$
Mass:	2.7 g	2.3–3.2 g	$n=4$

Vaseux Creek capture site.

MORPHOLOGICALLY SIMILAR SPECIES

Similar species that would be expected to co-occur with PREBLE'S SHREW are MERRIAM'S SHREW, CINEREUS SHREW and VAGRANT SHREW. Distinguishing PREBLE'S SHREW from them requires examination of cranial/dental traits or a genetic sample. They cannot be positively discriminated in the hand. MERRIAM'S SHREW is easily identified by the absence of a medial tine on its upper incisor and the shallow spaces between the denticles on its lower incisor. PREBLE'S SHREW is distinguished from CINEREUS SHREW and VAGRANT SHREW by its smaller cranial and dental measurements: skull length less than 14.5 millimetres, length of mandible less than 6.5 millimetres. VAGRANT SHREW also has smaller medial tines on the upper incisors that are usually positioned above the pigmented area, and its third upper unicuspid is usually smaller than the fourth.

NATURAL HISTORY

What little is known of this shrew's natural history in BC is limited to habitat descriptions. All capture sites were in shrub-steppe habitat at 312 to 1,724 metres elevation. The two sites near Vaseux Lake had a dense shrub cover (30 to 40 per cent) of ANTELOPE-BUSH 1.5 to 2.2 metres high. Most common herbaceous plants were CHEATGRASS, DIFFUSE KNAPWEED and NEEDLE-AND-THREAD GRASS. Both sites lacked tree cover, but scattered PONDEROSA PINES were nearby. The Mount Kobau site was in a patch of open grassland surrounded by sparse stands of DOUGLAS-FIR, BIG SAGEBRUSH and SNOWBERRY that formed a dense (80 per cent) shrub cover about 1 metre high. Various species of grasses

Heather Stewart at the Mount Kobau capture site.

formed the dominant ground cover. The Hamilton Commonage capture site was in upper grasslands characterized by sparse shrub cover of PASTURE SAGE and RABBIT-BRUSH. The habitat description for a PREBLE'S SHREW capture in the Columbia Basin of Washington is consistent with the Okanagan habitat. However, nine PREBLE'S SHREWS were captured in coniferous forest in the Blue Mountains of eastern Washington, raising the possibility of this species occurring in forested habitats in BC.

No other shrew species were captured at the Vaseux Lake sites, but DUSKY SHREWS and VAGRANT SHREWS were taken at the Mount Kobau study site. A single MERRIAM'S SHREW was the only shrew captured with PREBLE'S SHREW at the Washington site in the Columbia Basin.

Data on population numbers, movements and diet are lacking for this shrew from all parts of its range. Across its range PREBLE'S SHREW is generally described as occurring in low abundance. Sampling of 20 study sites in the south Okanagan with arthropod pitfall traps from 1993 to 1998 by Geoff Scudder yielded only three captures, the first captures for Canada reported by Nagorsen and colleagues. In 2013, 4,570 trap days in three grassland properties of the Nature Conservancy of Canada south of Richter Pass failed to capture any PREBLE'S SHREWS. Reproductive data are limited to a few individuals from Oregon captured in June and July. Five adult females were pregnant with three to six embryos. Four juvenile females showed no sign of reproductive activity. One young-of-the-year male had reached sexual maturity, but most juvenile males were not reproductively active.

CONSERVATION STATUS

This species is on the province's Red List; there has been no COSEWIC assessment. Nothing is known about its population size, population trends or range extent in the province. Two BC occurrences are in protected areas: the Vaseux-Bighorn National Wildlife Area and South Okanagan Grasslands Protected Area. Nonetheless, rarity and the loss of grassland and shrub-steppe habitat in BC from agriculture and urban development make it a species of concern. No information is available on the impacts of cattle grazing on this species. The BC capture sites are distant from the nearest known occurrence of PREBLE'S SHREW in Washington (Douglas County), but we attribute this to a lack of shrew sampling in northern parts of the state rather than a discontinuity in suitable habitat.

SYSTEMATICS AND TAXONOMY

PREBLE'S SHREW is a member of the *Sorex cinereus* group that comprises 13 related species found in North America and Siberia. A study of mitochondrial and nuclear DNA confirmed that the 13 species fall into two major lineages—Beringian and Southern. PREBLE'S SHREW is a member of the Beringian lineage, with MT. LYELL SHREW and PRAIRIE SHREW appearing to be its closest relatives. Nonetheless, the taxonomy of PREBLE'S SHREW is unresolved. No subspecies are recognized, but the DNA data suggest that populations of PREBLE'S SHREW in the Rocky Mountains are sufficiently distinct that they may be an incipient species.

REMARKS

Similar to MERRIAM'S SHREW, the first evidence of this species in the province and Canada were the three fatalities found in 1994 to 1995 in arthropod traps set by entomologist Geoff Scudder in the Okanagan. David Huggard, a graduate student then, tentatively identified them as PREBLE'S SHREW. The three specimens were subsequently deposited with the Royal BC Museum as vouchers, and we confirmed their species identification. The occurrence in the Hamilton Commonage was an incidental capture in 1993 by Astrid van Woudenberg as part of a study on grazing impacts on vertebrate communities.

The species name *preblei* is an eponym for Edward Alexander Preble, a naturalist with the US Biological Survey who carried out extensive fieldwork on birds and mammals in western North America, including an 1897 expedition that covered much of the BC coast and a 1913 expedition over a vast area in northern BC.

REFERENCES

Armstrong (1957); Carraway and Verts (1999); Cornely, Carraway and Verts (1992); Gitzen et al. (2009); Hope et al. (2012); Nagorsen et al. (2001).

Olympic Shrew

Sorex rohweri

Other common name: Rohwer's Shrew

DESCRIPTION

OLYMPIC SHREW is a small shrew with dorsal fur that is brown in summer pelage and dark grey in winter. The tail is strongly bicoloured with a paler ventral side. The third and fourth digits of the hind foot have five pairs of toe pads.

Cranial/dental traits: 32 teeth; incisors 1/1, unicuspids 5/1, premolars 1/1, molars 3/3; upper incisor with a minute to small medial tine positioned above, at or below the pigment edge; in some individuals the medial tine is not evident except for a flattening of each incisor where they are in contact; first upper unicuspid weakly or not pigmented to the cingulum; third upper unicuspid slightly larger, equal or slightly shorter than the fourth; lower incisor has three denticles with continuous pigmentation on the labial side covering one denticle, with the pigment on the lingual side extending to the first or second denticle; a small to large postmandibular foramen is present in both dentaries.

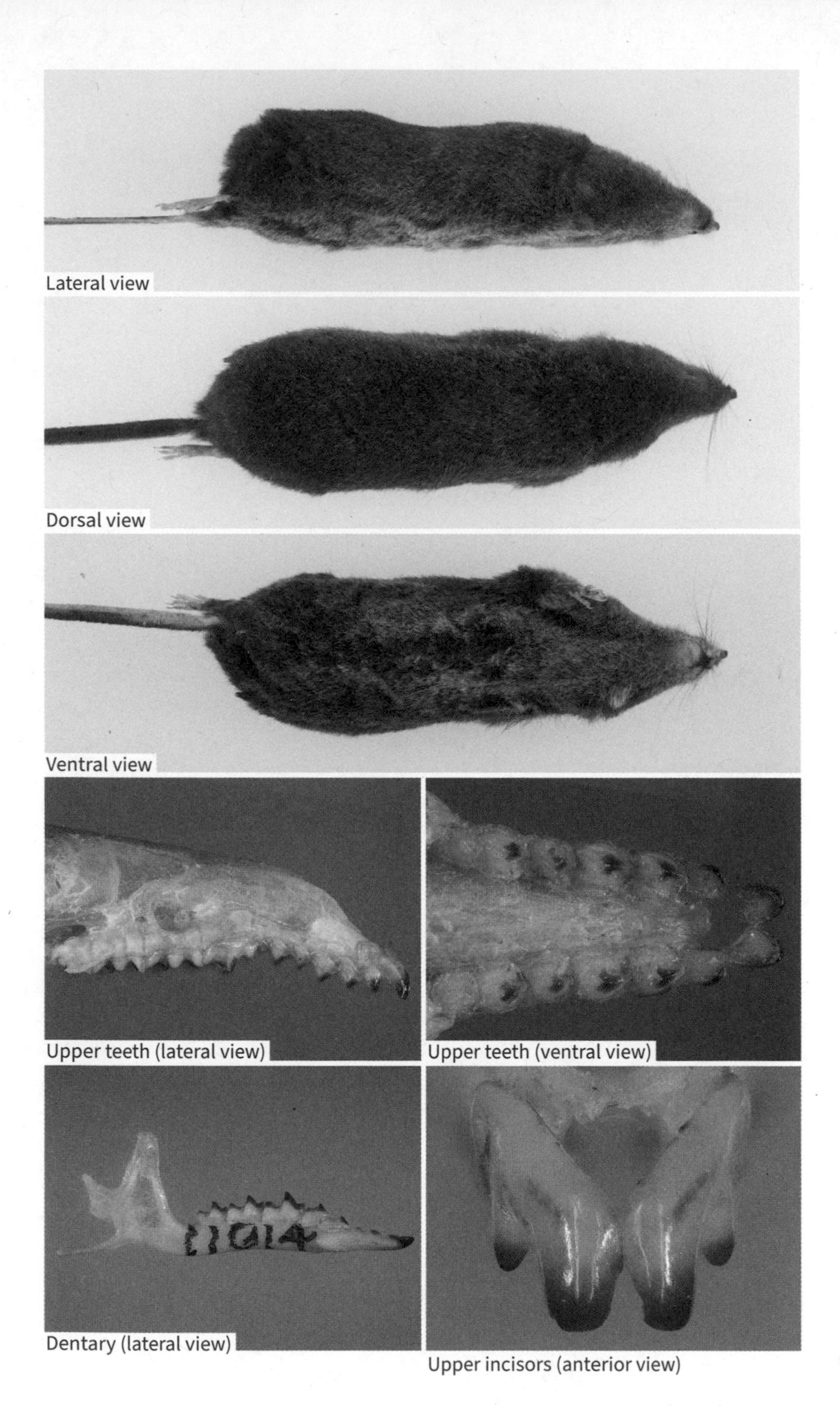

Lateral view

Dorsal view

Ventral view

Upper teeth (lateral view)

Upper teeth (ventral view)

Dentary (lateral view)

Upper incisors (anterior view)

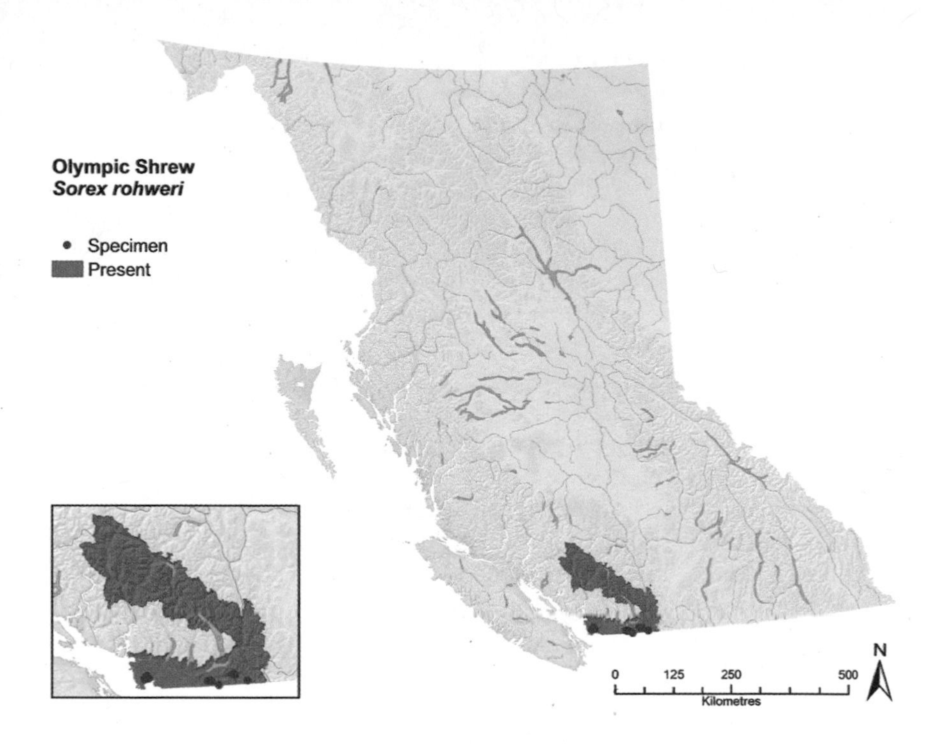

DISTRIBUTION

OLYMPIC SHREW is a coastal species, and its known occurrences range from western Oregon, the Olympic Peninsula and Cascade Mountains of Washington to the lower Fraser River basin in extreme southwestern BC. There are 13 known occurrences in the province that are confined to a small range extent of about 360 square kilometres on the south side of the Fraser River extending from Burns Bog in Delta and to Chilliwack Lake in the Chilliwack valley. Eastern limits of its range are unknown, but it may reach the Skagit valley.

MEASUREMENTS

	Mean	Range	Sample size
Total length:	101 mm	91–109 mm	$n = 14$
Tail vertebrae:	43 mm	32–50 mm	$n = 17$
Hind foot:	11 mm	8–13 mm	$n = 17$
Mass:	4.4 g	2.5–8.0 g	$n = 6$

MORPHOLOGICALLY SIMILAR SPECIES

Of the six shrews that co-occur with OLYMPIC SHREW in BC, most similar are CINEREUS SHREW and VAGRANT SHREW. Of the 18 known museum specimens of OLYMPIC SHREW from the province, 16 were originally identified on their labels as CINEREUS SHREWS and two as VAGRANT SHREWS. Discriminating these three species from external morphological traits is not reliable as they overlap extensively in body measurements and show few clear differences in pelage colour or other traits. OLYMPIC SHREWS may differ from CINEREUS SHREWS and VAGRANT SHREWS in having five toe pads on the middle digits of their hind feet and a bicoloured tail, but more samples are required to determine the extent of variation in these traits. Positive identification requires examination of either cranial/dental traits or genetic samples. CINEREUS SHREW differs from the OLYMPIC SHREW in having teeth more darkly pigmented, a small to large medial tine on its upper incisor that is situated well below the edge of the pigment, and a strongly pigmented ridge on the first upper unicuspid that extends to the cingulum. VAGRANT SHREW has a medium-sized tine on the upper incisors, and all three denticles on its lower incisor are within the pigmented area.

NATURAL HISTORY

Data on population numbers, movements and diet are lacking for this shrew from all parts of its range. Reproductive data for BC is limited to a single female captured May 7 in Burns Bog that had five embryos. What little is known of the OLYMPIC SHREW's natural history is mostly based on habitat descriptions. In BC, it appears to be associated with mixed or coniferous forests of various seral stages. Captures in Burns Bog were in or in close proximity to forests of LODGEPOLE PINE, RED ALDER, WESTERN HEMLOCK, WESTERN REDCEDAR or birch. Immature or old-growth DOUGLAS-FIR forest was mentioned in habitat descriptions for specimens captured in the Chilliwack valley. Although most occurrences are below 700 metres elevation, three museum specimens (originally identified as CINEREUS SHREW until our study) taken in 1896 by Allan Brooks at about 1,400 metres elevation on Liumchen Mountain suggest that populations extend into the subalpine zone of the Cascade Mountains in BC. On the eastern slope of the Cascade Mountains near Snoqualmie Pass in Washington, 41 OLYMPIC SHREWS were captured in Jordan Ryckman's study sites situated at 700 to 800 metres elevation.

CONSERVATION STATUS

This species is on the province's Red List; it has not been assessed by COSEWIC. Nothing is known about its population size or trends. However, the available habitat information suggests that OLYMPIC SHREW is forest dependent; therefore, it may be impacted by the loss of forests from agriculture and urban development

Nancy Nagorsen at capture site in Burns Bog, BC.

that has occurred over the past 150 years in the Fraser River valley. Suitable habitat in this region now consists of isolated patches of forest in Burns Bog, Sumas Mountain, Vedder Mountain and the Chilliwack valley and montane forests in the Cascade Mountains. In the past 25 years, all captures of the OLYMPIC SHREW were from Burns Bog, a 3,500-hectare peat bog with heath and forest ecosystems surrounded by the densely urbanized landscape of the municipality of Delta. However, this may reflect the intensive sampling done in Burns Bog for environmental assessments associated with the construction of Highway 17. A conservation assessment of this species will require more research to confirm the current distribution, habitat requirements, and connectivity of BC populations with those of populations in nearby Washington State.

SYSTEMATICS AND TAXONOMY

The OLYMPIC SHREW was described and named in 2007 by Robert Rausch and colleagues based on a study of recent captures and historical museum specimens from the Olympic Peninsula of Washington that were originally misidentified as CINEREUS SHREWS. A subsequent genetic study of mitochondrial and nuclear DNA confirmed its species status and revealed that its nearest relatives are members of the *Sorex cinereus* group. Despite morphological similarity, DNA sequence data suggest that the OLYMPIC SHREW diverged from the *Sorex cinereus* group some 850,000 years ago. There has been no study of geographic variation in morphology or genetics, and no subspecies are recognized.

REMARKS

Given the long history of small-mammal surveys and research in the Pacific Northwest, the discovery of a new shrew species in 2007 is remarkable. It demonstrates the importance of museum voucher specimens for revealing cryptic species. The presence of this species in coastal Oregon was only confirmed in 2016 when Neal Woodman and Robert Fisher found that museum specimens housed in the collections of the National Museum of Natural History (Smithsonian Institution) originally identified as CINEREUS SHREWS were OLYMPIC SHREWS. No research or inventory has been done on the OLYMPIC SHREW in BC since our study in 2009. Virtually every aspect of this shrew's biology is in need of study.

With a range encompassing coastal Oregon, Olympic Mountains, Cascade Mountains and the lower Fraser River basin, the common name OLYMPIC SHREW seems inappropriate. The species name *rohweri* is an eponym for Sievert A. Rohwer, a former director of the Burke Museum at the University of Washington.

REFERENCES

Brooks (1902); Hope et al. (2012); Nagorsen and Panter (2009); Rausch, Feagin and Rausch (2007); Ryckman (2020); Woodman and Fisher (2016).

Trowbridge's Shrew *Sorex trowbridgii*

DESCRIPTION

TROWBRIDGE'S SHREW is a medium-sized shrew with a dark-grey or sooty dorsal pelage and paler underparts. The long tail is distinctly bicoloured with a dark-grey dorsal surface and nearly white underside. The winter pelage tends to be more grey than the summer pelage. The moult to summer pelage begins in late April and is complete by June. The moult to winter pelage begins in September.

Cranial/dental traits: 32 teeth; incisors 1/1, unicuspids 5/1, premolars 1/1, molars 3/3; upper incisor with a small to large medial tine partly in the upper edge, or slightly above the pigmented region; third upper unicuspid shorter than the fourth; a weakly pigmented ridge on the upper unicuspids does not extend to the cingulum and is separated from the cingulum by a longitudinal groove or unpigmented area; lower incisor has three denticles with continuous pigmentation on the labial side including two or three denticles, with a lingual pigmented area extending to the second denticle; medium to large postmandibular foramen in both dentaries that is usually confluent with the mandibular foramen.

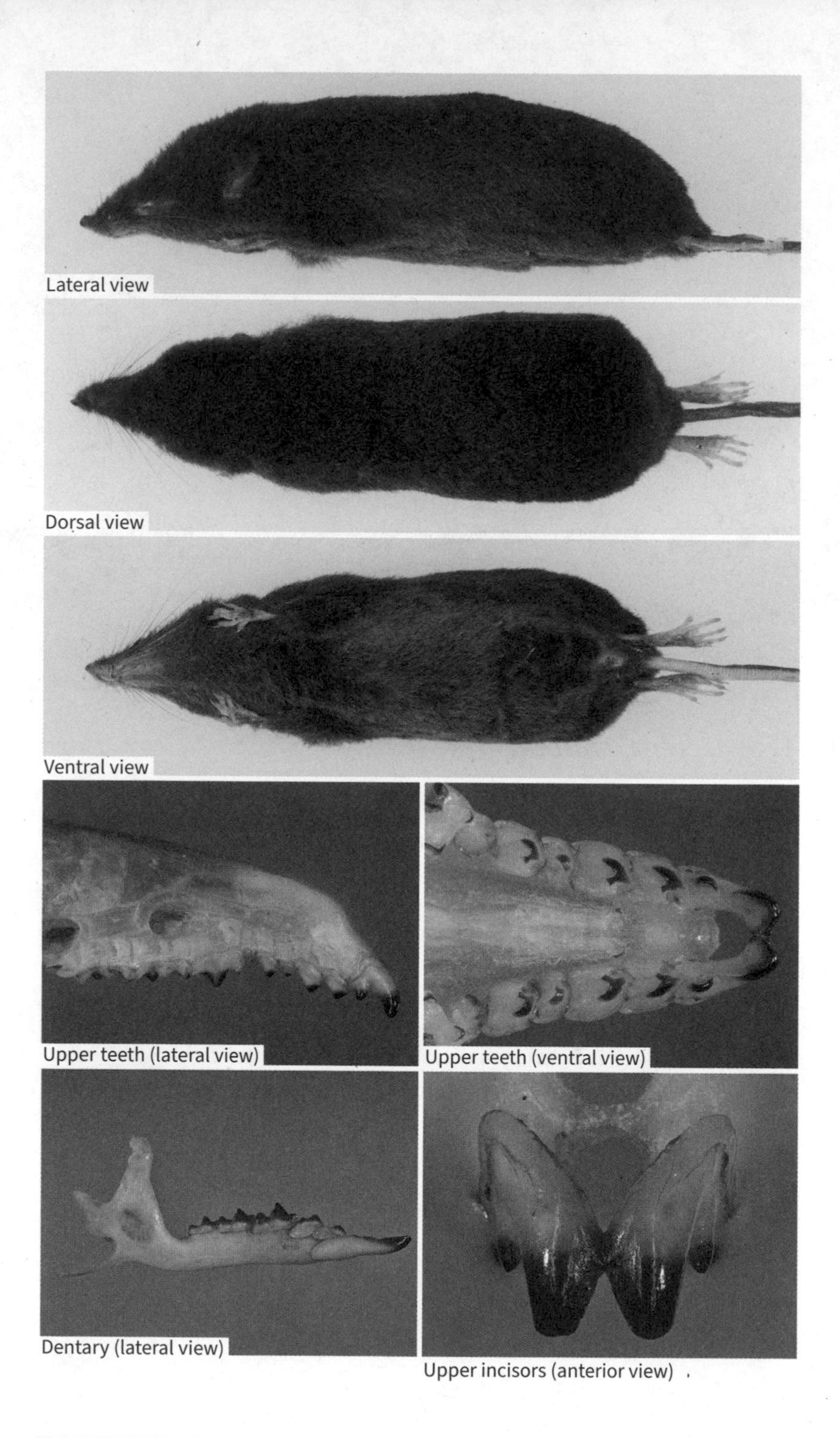

Upper incisors (anterior view)

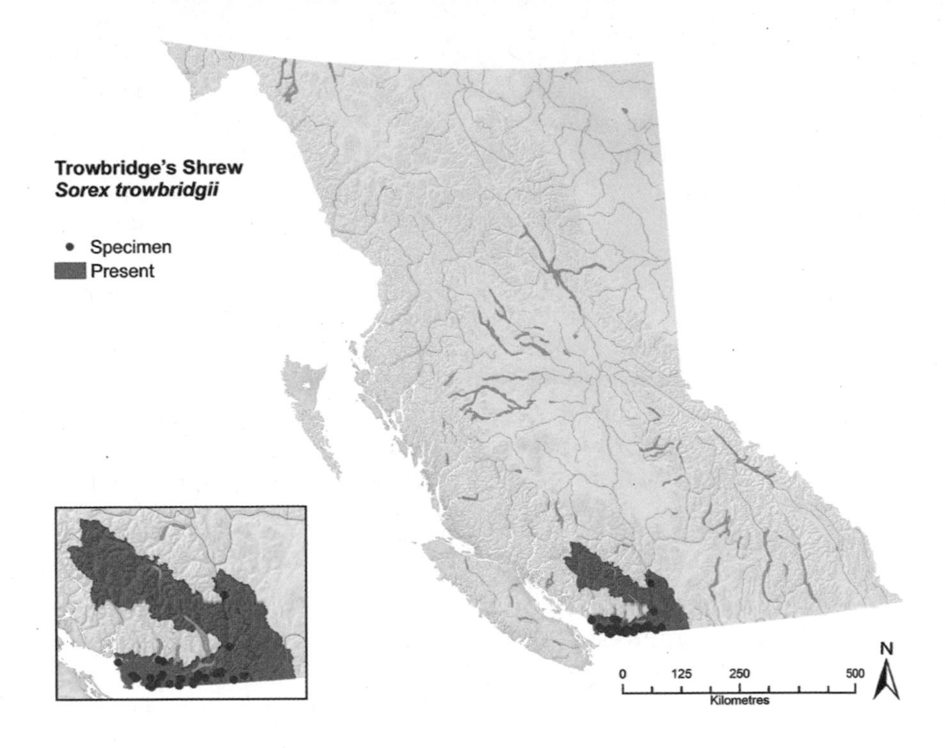

DISTRIBUTION

TROWBRIDGE'S SHREW ranges from California through western Oregon and Washington to extreme southwestern BC, where it is restricted to the lower Fraser River valley. Easternmost occurrences are Chilliwack Lake and Hope. There is a historical museum specimen in the Canadian Museum of Nature that was labelled from the "second Summit, Skagit River," but we were unable determine the precise location of this site. Northernmost confirmed occurrence is 10 kilometres north of Boston Bar east of the Fraser River. There are only three confirmed records north of the Fraser River—Point Grey in Vancouver, Maple Ridge and Harrison Lake.

MEASUREMENTS

	Mean	Range	Sample size
Total length:	118 mm	101–130 mm	$n = 198$
Tail vertebrae:	55 mm	45–62 mm	$n = 197$
Hind foot:	13 mm	10–15 mm	$n = 197$
Mass:	5.5 g	3.6–8.0 g	$n = 26$

MORPHOLOGICALLY SIMILAR SPECIES

Within its BC range, this species could be confused with OLYMPIC SHREW, CINEREUS SHREW, DUSKY SHREW, PACIFIC SHREW or VAGRANT SHREW. OLYMPIC SHREW is smaller, but in winter pelage it resembles TROWBRIDGE'S SHREW with a dark-grey pelage and bicoloured tail. A sharply bicoloured tail, white underneath and with sooty-grey fur, discriminates TROWBRIDGE'S SHREW from CINEREUS SHREW, DUSKY SHREW, PACIFIC SHREW and VAGRANT SHREW. Cranial/dental traits are more reliable for identification. CINEREUS SHREW and OLYMPIC SHREW usually have a third upper unicuspid tooth equal or taller than the fourth and a shorter skull (skull length less than 16.0 millimetres). OLYMPIC SHREW has only one denticle included in the pigmented area of the lower incisor. PACIFIC SHREW and VAGRANT SHREW have a third unicuspid smaller than the fourth similar to TROWBRIDGE'S SHREW, but they have a pigmented ridge on their upper unicuspid teeth that extends to the cingulum without a longitudinal groove. The width across the third upper molars of VAGRANT SHREW is less than 4.1 millimetres.

NATURAL HISTORY

In coastal regions, TROWBRIDGE'S SHREW is most common in dry mixed and coniferous forests with rich soil and abundant decaying wood and litter on the forest floor. Nevertheless, it has broad habitat requirements occupying wet forests, riparian habitats and ravines. It generally avoids damp marshy areas with saturated soil. TROWBRIDGE'S SHREW constructs tunnels in the humus layer of the forest litter, which may explain why it prefers habitats with dry, loose soil and deep litter, and avoids wet areas with a high water table. Captive animals constructed interconnected tunnels up to 30 centimetres long within the humus layer. Several studies in coastal habitats of the United States have shown that it occupies a wide range of forest stands, from recent clearcuts to old-growth forests. In BC, TROWBRIDGE'S SHREW has been found at elevations up to 640 metres, but most populations are at lower elevations. Gustavo Zuleta and Carlos Galindo-Leal found it to be widespread and common in various habitats south of the Fraser River, with the highest captures in dry, mixed forests of RED ALDER, WESTERN HEMLOCK, WESTERN REDCEDAR and BIGLEAF MAPLE. It was found in forests of various ages, but only in young forests with extensive canopy cover. In Burns Bog, where five species of shrew occur, Mark Fraker and colleagues captured TROWBRIDGE'S SHREW in pine woodland and mixed deciduous forest sites; it was not captured in various wetland habitats of the bog.

Although there are no estimates of absolute numbers, results from field studies suggest that TROWBRIDGE'S SHREW is the most common shrew in many coastal forest habitats. A study in different-aged forest stands dominated by WESTERN HEMLOCK on the Olympic Peninsula of Washington found it to be the most common small mammal occurring in all study sites, but greatest captures were in old-growth forests. United States Forest Service surveys in the Cascade Mountains of Oregon and Washington revealed that it was the dominant small mammal in DOUGLAS-FIR forests. Similarly, it was the most commonly captured shrew in Gustavo Zuleta and Carlos Galindo-Leal's study area in the lower Fraser River valley of BC.

Captive TROWBRIDGE'S SHREWS demonstrate short periods of activity at regular intervals over a 24-hour period; they are active about 39 per cent of the time, mostly after dark. Animals in breeding condition are more active than young animals or non-breeding adults.

A diverse assortment of soil-dwelling and surface-dwelling invertebrates are eaten. Centipedes, spiders, slugs, snails, beetles and other adult and larval insects are the major prey eaten in western Oregon, with centipedes found in about one-third of the TROWBRIDGE'S SHREW stomachs analyzed. Although these are minor food items, this shrew also consumes plant material. Fungi, including underground fungi or truffles, have been found in the stomachs of a few individuals, and several researchers noted this shrew's tendency to feed on seeds from coniferous trees such as DOUGLAS-FIR. Captive TROWBRIDGE'S SHREWS readily consumed seeds from various herbs, shrubs and trees. Hoarding behaviour was common among captives; they transported seeds a considerable distance from the feeding trays and then buried them for later consumption.

In the Sierra Nevada of California, TROWBRIDGE'S SHREW has an earlier breeding season than other shrew species. Reproductive activity begins in February, with females reaching sexual maturity several weeks after males. Pregnant females were observed from February to late May, with most reproductive activity finished by June. In western Oregon, reproduction begins in February, and 97 per cent of the females captured in May were either pregnant or nursing, and few breeding females were found after June. The number of embryos ranges from three to six, with four or five most common. Females are capable of producing at least two litters during a breeding season. The meagre data for BC taken from museum specimens suggest a similar breeding season. Dates for three pregnant females were March 25, May 5 and May 9 with embryo counts from three to six. Two females captured on May 5 and June 3 were nursing.

Owls, especially the BARN OWL and BARRED OWL, are probably the major predator of TROWBRIDGE'S SHREW in BC.

CONSERVATION STATUS

TROWBRIDGE'S SHREW is on the provincial Blue List; there has been no COSEWIC assessment. Although abundant in forested habitats, local populations may be at long-term risk because much of its range coincides with a region undergoing rapid urban growth and habitat loss. Gustavo Zuleta and Carlos Galindo-Leal noted that much of this species's forest habitat on the south side of the Fraser River is fragmented from development.

SYSTEMATICS AND TAXONOMY

In a phylogenetic study using allozymes as genetic markers, Sarah George found that TROWBRIDGE'S SHREW was a member of a distinct clade of North American shrews that included ARIZONA SHREW and MERRIAM'S SHREW. Molecular genetics based on mitochondrial DNA support the association of TROWBRIDGE'S SHREW with a lineage comprising ARIZONA SHREW and several Central American shrew species. This distinct and old lineage diverged from other North American shrews in the Miocene epoch some 10 to 5 million years ago.

Five subspecies based on morphology are recognized; one occurs in BC. A morphometric study of skull and dentary measurements suggested that much of the variation among the five races may be the result of a south-to-north trend of decreasing size. A genetic study is required to evaluate the validity of the five races.

- *Sorex trowbridgii trowbridgii* Baird. Ranges from northwestern California to BC.

REMARKS

The distribution and status of this shrew on the north side of the Fraser River are an enigma. Only a few historical museum specimens exist taken at three locations. Of the 37 study sites north of the Fraser River sampled by Gustavo Zuleta and Carlos Galindo-Leal, they captured TROWBRIDGE'S SHREW at only one location.

The species name *trowbridgii* is an eponym named for William P. Trowbridge, an American military officer and naturalist who collected the first TROWBRIDGE'S SHREW specimen in 1855 at Astoria, Oregon.

REFERENCES

Aubry, Crites and West (1991); Carey and Johnson (1995); Carraway (1987); Carraway and Verts (2005); Fraker, Bianchini and Robertson (1999); Gashwiler (1976); George (1988, 1989); Jameson (1955); Lehmkuhl, Peffer and O'Connell (2008); Maldonado et al. (2015); Rust (1978); Terry (1978, 1981); Whitaker and Maser (1976); Zuleta and Galindo-Leal (1994).

Tundra Shrew *Sorex tundrensis*

DESCRIPTION

TUNDRA SHREW has an attractive and distinctive pelage. In summer, adults are tricoloured with a brown dorsal stripe, greyish-pale-brown sides and grey underparts. The dark dorsal fur contrasts strikingly with the paler sides, creating a saddlebacked pattern. Immature animals have a greyish back that is not as strongly defined from the paler sides as it is in adults. The tail is indistinctly bicoloured, brown above and buff underneath; the feet are light brown. In winter pelage, the fur is bicoloured: the grey sides and belly contrast with a brown back.

Cranial/dental traits: 32 teeth; incisors 1/1, unicuspids 5/1, premolars 1/1, molars 3/3; upper incisor with small medial tine positioned well within the pigmented area; third upper unicuspid taller than the fourth; edge of pigment on anterior face of upper incisor sits above the pigment on its lateral cusp; hypocone on first upper molar usually pigmented; cusps of upper unicuspid teeth not connected to the cingulum by a pigmented ridge; lower incisor has three denticles with continuous pigmentation on the labial side covering three denticles, with pigment on lingual side only extending to first denticle; medium to large postmandibular foramen in both dentaries.

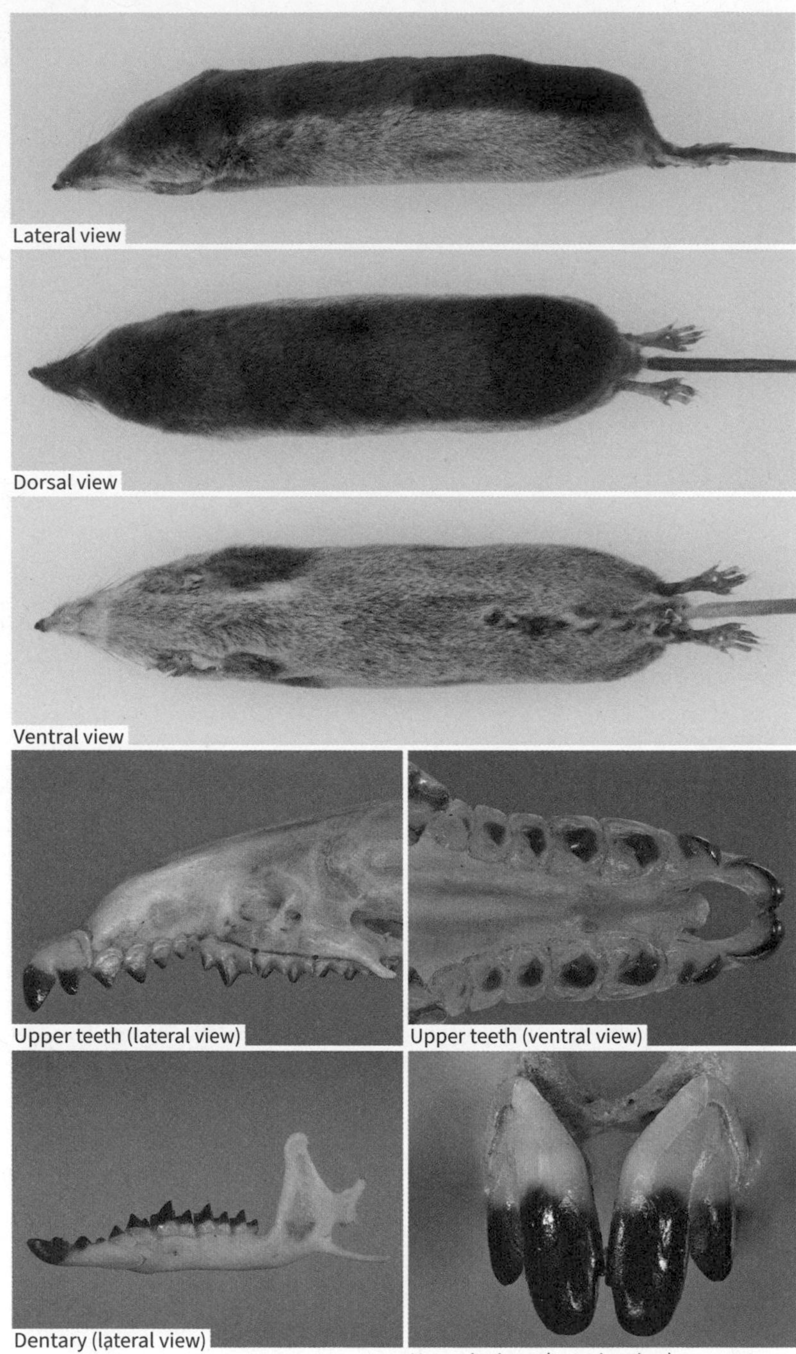
Lateral view
Dorsal view
Ventral view
Upper teeth (lateral view)
Upper teeth (ventral view)
Dentary (lateral view)
Upper incisors (anterior view)

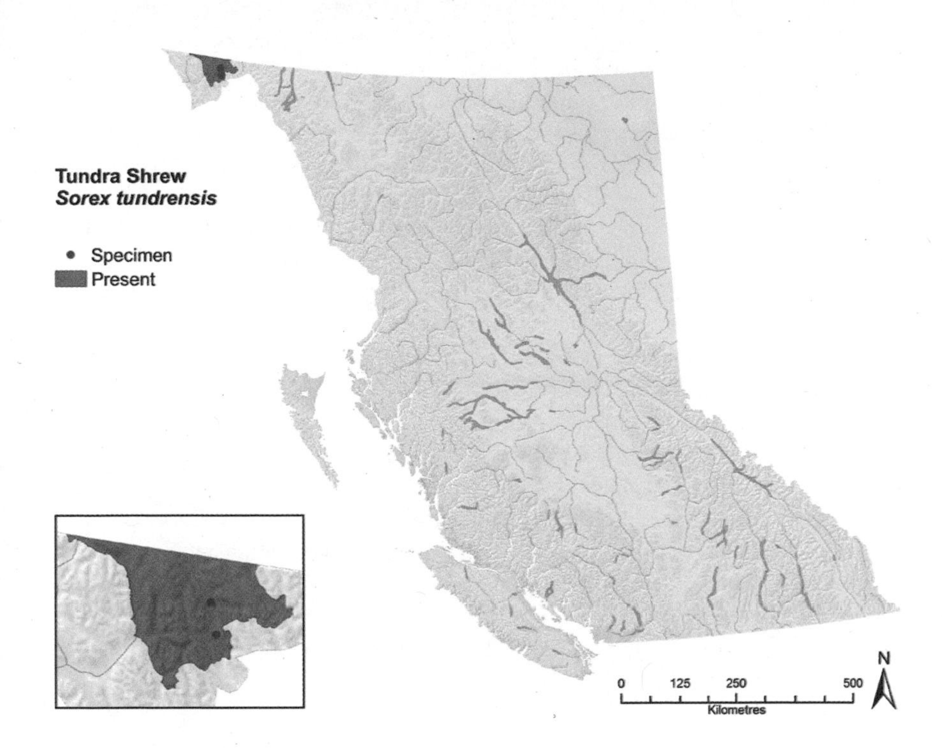

DISTRIBUTION

An Arctic species, TUNDRA SHREW inhabits Eurasia and North America—Alaska, the northern Yukon, the Mackenzie delta region of the Northwest Territories, and northwestern BC. Confined to the extreme northwestern corner of BC, it has one of the most restricted distributions of any mammal in the province. There are only three known occurrences: one east of the Haines Highway in the Chilkat Pass and two west of Kelsall Lake close to the Haines Highway in Tatshenshini-Alsek Provincial Park. They represent the southernmost occurrences for this shrew in North America.

MEASUREMENTS

Measurements are based on specimens from the Yukon and Northwest Territories.

	Mean	Range	Sample size
Total length:	97 mm	84–115 mm	*n* = 31
Tail vertebrae:	31 mm	25–37 mm	*n* = 35
Hind foot:	13 mm	11–15 mm	*n* = 34
Mass:	6.8 g	4.9–10.0 g	*n* = 29

MORPHOLOGICALLY SIMILAR SPECIES

Our only shrew with similar markings and dental traits is the ARCTIC SHREW. Found in the northeastern part of the province east of the Rocky Mountains, the ARCTIC SHREW's BC range is not known to overlap with that of TUNDRA SHREW. The only shrews that could be confused with TUNDRA SHREW are CINEREUS SHREW and DUSKY SHREW. Neither has a prominent tricoloured dorsal pelage. Both species have an upper first unicuspid pigmented to the cingulum and lack pigment on the hypocone of the first upper molar. DUSKY SHREW has a third upper unicuspid that is shorter than the fourth, and larger medial tines on the upper incisors. CINEREUS SHREW shares a similar pattern to TUNDRA SHREW in the relative size of its unicuspid teeth but has a smaller skull (skull length less than 16.0 millimetres).

NATURAL HISTORY

In Alaska and Arctic regions of northwestern Canada, TUNDRA SHREW lives in tundra habitats with DWARF BIRCH and willow thickets or taiga forests. The isolated BC population is situated on the leeward side of the St. Elias Mountains, a region with a distinctly subarctic climate despite its proximity to the Pacific Ocean. Of the three known sites in the province, one is in an alpine willow community bordering a stream (1,037 metres elevation); the others are in subalpine willow and grass communities (885 metres elevation). Characteristic plants in these communities include willows, DWARF BIRCH, MOUNTAIN SAGEWORT, NORTHERN GRASS-OF-PARNASSUS, UNALASKA PAINTBRUSH, horsetails and various grasses and sedges.

Capture site at Datlasaka Flats, near Haines Highway.

TUNDRA SHREW behaviour has not been studied, and what little is known about its biology is based on anecdotal accounts. It appears to be active at all hours. In summer, it may use well-defined runways constructed by other mammals. Its movements and home range have not been determined. High numbers have been reported in the Arctic, but there are no data on populations in BC habitats.

Data on food habits are limited to Alaska, where insect larvae, earthworms and some plant material were identified in stomach remains. No reproductive data exist for the BC population. In the Arctic, the breeding season extends from April to September. Males become sexually active by May; pregnant or nursing females have been reported from May to early September. Embryo counts range from 8 to 12. Females may produce two litters in the breeding season and may breed in their first summer.

CONSERVATION STATUS

TUNDRA SHREW has not been assessed by COSEWIC because it is not at risk in most of its Canadian range. Although two occurrences are within the boundaries of Tatshenshini-Alsek Provincial Park, this shrew appears on the BC Red List because of rarity and its isolation from populations in Alaska and Yukon. If it depends on tundra-like habitat, a possible long-term threat is climate change and associated habitat changes such as increasing shrub growth along the Haines Highway area.

SYSTEMATICS AND TAXONOMY

TUNDRA SHREW is a member of the *Sorex araneus* group, comprising about a dozen species most of which inhabit Eurasia. It is the only member of the group found in both North America and Eurasia. It was classified as a subspecies of the ARCTIC SHREW in some early literature, but DNA studies have shown that they are distantly related species in the *Sorex araneus* group, with the closest relative of TUNDRA SHREW being the TIEN SHAN SHREW of China and Kazakhstan. A study based on mitochondrial DNA revealed seven distinct genetic clades of TUNDRA SHREW with six in Eurasia and one in North America. Genetic divergence of the North American clade from Eurasian populations occurred 120,000 years ago consistent with it crossing the Bering Land Bridge and persisting in northwestern North America throughout the last glaciation. A number of subspecies are recognized, but they are inconsistent with the genetic data.

- *Sorex tundrensis tundrensis* Merriam. A widespread race that includes the North American populations and populations in eastern Siberia. However, the described range of this subspecies is not supported by the mitochondrial DNA study that found North American and eastern Siberian populations falling into different genetic clades.

REMARKS

TUNDRA SHREW was discovered in BC in the summers of 1978 and 1979, when Donald Jones captured five during surveys as part of his graduate thesis study on the diet of RED FOX along the Haines Highway. In 1983, biologists from the Royal BC Museum including the senior author carried out intensive small-mammal surveys including pitfall trapping in subalpine habitats at Tats Lake, west of the Tatshenshini River. They captured large numbers of CINEREUS SHREWS and DUSKY SHREWS but no TUNDRA SHREWS, suggesting that it may be restricted to tundra habitats along the Haines Highway. TUNDRA SHREW has not been found in southeastern Alaska or southern Yukon, and the BC captures are about 500 kilometres south of the nearest known occurrences. They may represent a small relict population that was left isolated in tundra habitats that persisted along the Haines Highway after the last ice age.

No BC field surveys for this species have been done in the past 41 years. We recommend a focused research study in this remote corner of BC to determine its population, habitat associations, distribution and conservation status.

The species name *tundrensis* means "of the tundra."

REFERENCES

Bee and Hall (1956); Hope et al. (2011); Junge, Hoffmann and Debry (1983); Mackiewicz et al. (2017); Martell and Pearson (1978); Nagorsen and Jones (1981); Quay (1951); Youngman (1975).

Vagrant Shrew

Sorex vagrans

Other common name: Wandering Shrew

DESCRIPTION

VAGRANT SHREW is a medium-sized shrew with a dorsal pelage that varies from greyish to dark brown in Interior populations to nearly black in coastal populations. The ventral fur is brown to grey. The winter pelage is longer and darker than the summer pelage. In southwestern BC, the moult to winter pelage begins in mid-September; the moult to summer pelage begins in early March and is complete by the first week of April. Females acquire their summer pelage before pregnancy. Prominent flank glands appear in breeding males in January or February. The tail is not distinctly bicoloured in adults. There are no more than four pairs of toe pads on the second to fifth digits of the hind feet.

Cranial/dental traits: 32 teeth; incisors 1/1, unicuspids 5/1, premolars 1/1, molars 3/3; upper incisor with a medium-sized medial tine near the top edge of pigmented area; medial tines often separated from pigmented area by a pale-coloured gap; in some coastal populations, pigmentation pale on the upper incisor and medial tine; third upper unicuspid equal or shorter than the fourth; upper unicuspids with a weak to strongly pigmented ridge that extends to the cingulum; lower incisor has three denticles with pigment on the labial side continuous to the third denticle, with lingual pigmented area extending to the second or third denticle; postmandibular foramen present in only 15 per cent of specimens examined, and if present, tiny or small and in only one dentary.

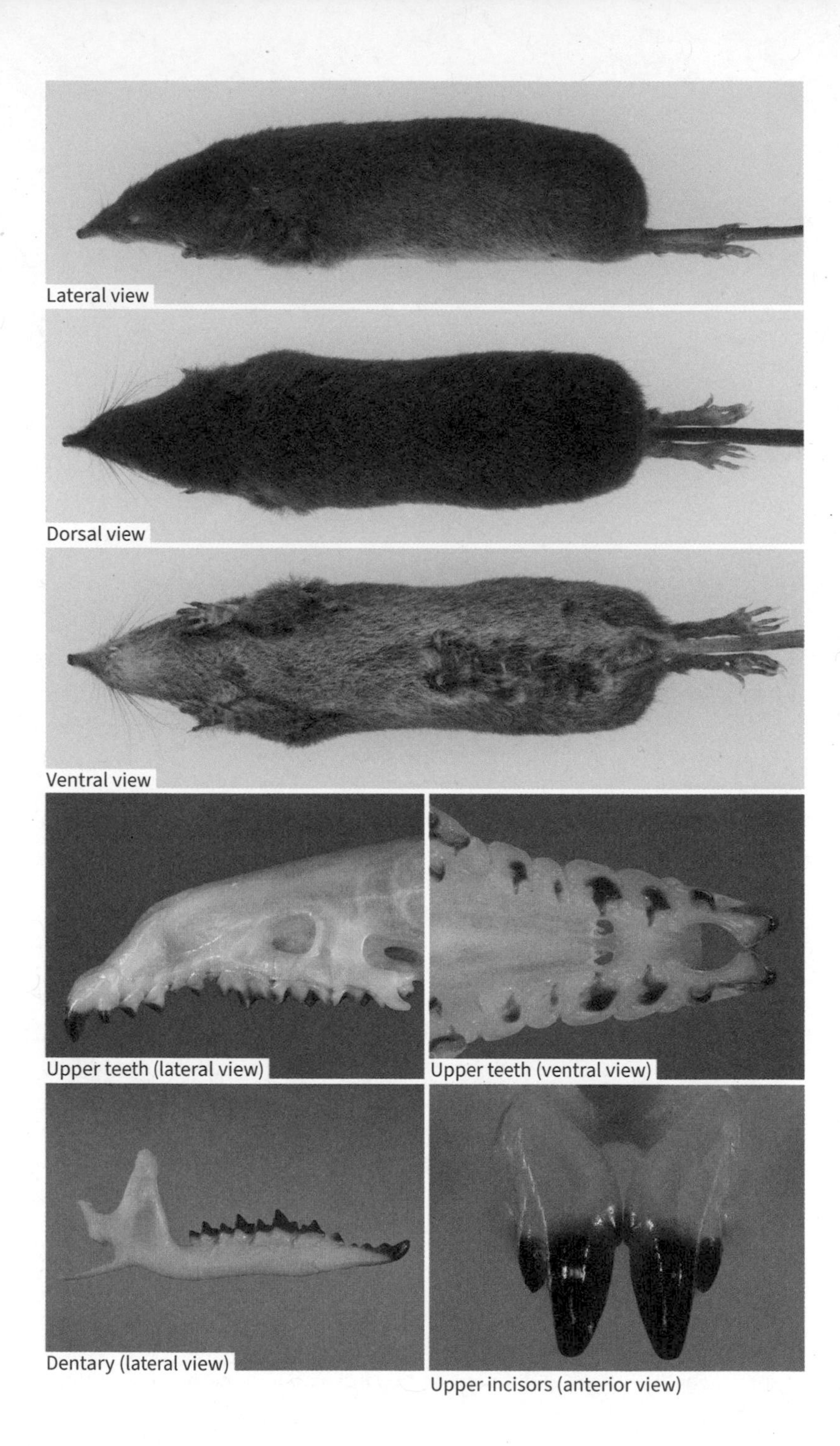
Lateral view
Dorsal view
Ventral view
Upper teeth (lateral view)
Upper teeth (ventral view)
Dentary (lateral view)
Upper incisors (anterior view)

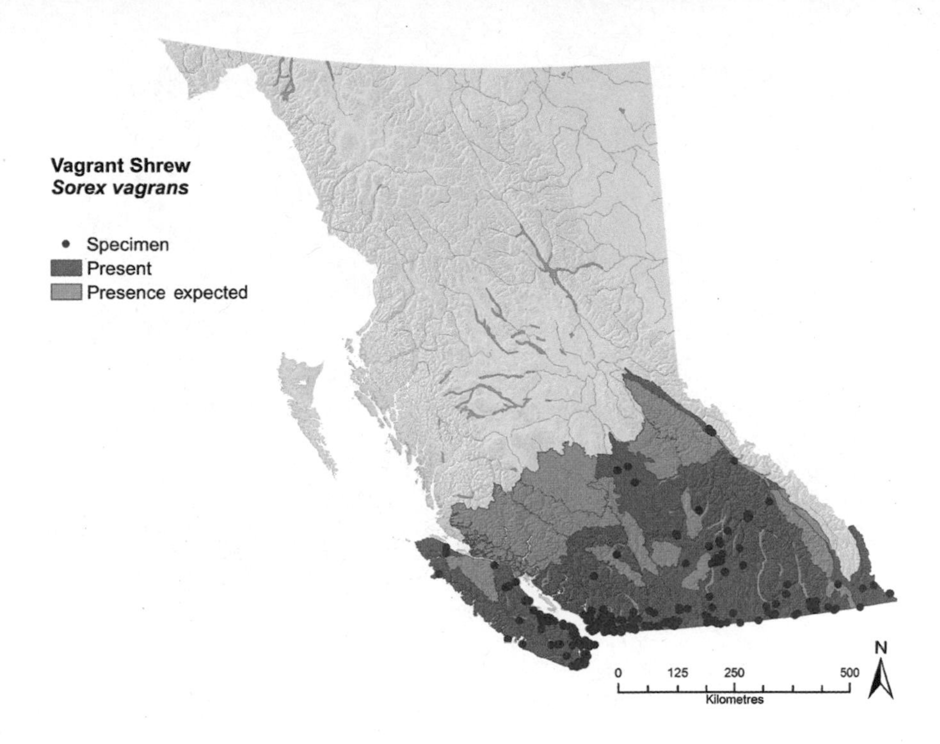

DISTRIBUTION

VAGRANT SHREW ranges across much of the western United States from California and the southern Rocky Mountains north to BC and extreme southwestern Alberta. It is found across the southern mainland of the province as far north as Alta Lake in the Coast Mountains, and Williams Lake and the Kinbasket Reservoir near Valemount in the Interior. Eastern limits are the Rocky Mountain Trench and east of the Flathead River near the continental divide (Sage Creek) in the Rocky Mountains. This shrew also inhabits numerous coastal islands including Vancouver Island and the southern Gulf Islands. Northern limits of this species's range on the coastal mainland, the Interior Plateau and along the continental divide of the Rocky Mountains are poorly defined.

MEASUREMENTS

	Mean	Range	Sample size
Total length:	104 mm	85–126 mm	*n* = 799
Tail vertebrae:	43 mm	32–60 mm	*n* = 797
Hind foot:	12 mm	9–15 mm	*n* = 797
Mass—females:	5.1 g	3.0–10.0 g	*n* = 336
Mass—males:	5.5 g	3.0–9.0 g	*n* = 371

Southern Rocky Mountains east of the Flathead River where VAGRANT SHREW ranges into the subalpine.

MORPHOLOGICALLY SIMILAR SPECIES

Species that co-occur with VAGRANT SHREW that may cause misidentification are MERRIAM'S SHREW, CINEREUS SHREW, OLYMPIC SHREW, TROWBRIDGE'S SHREW, DUSKY SHREW and PACIFIC SHREW. MERRIAM'S SHREW is easily distinguished by the absence of medial tines on its upper incisor, and some individuals have a white-tipped tail. CINEREUS SHREW has six pairs of toe pads on the third, fourth and fifth digits of the hind foot, and a third upper unicuspid taller than or equal in size to the fourth. TROWBRIDGE'S SHREW has a third upper unicuspid shorter than the fourth, but it has a strongly bicoloured tail and a pigmented ridge on the upper unicuspid teeth separated from the cingulum by a longitudinal groove. OLYMPIC SHREW, DUSKY SHREW and PACIFIC SHREW closely resemble VAGRANT SHREW in external and cranial/dental traits, but VAGRANT SHREW differs from these three species in having no more than four toe pads on the digits of its hind feet, a medial tine on the upper incisor positioned above the pigmented region bordered by a pale band, and a width across second upper molars less than 4.5 millimetres.

NATURAL HISTORY

Although VAGRANT SHREW inhabits a wide range of habitats, it is typically associated with moist forests, open patches in forests, and swamps, bogs and grassy meadows. In southwestern BC, where detailed habitat data are available from Myrnal Hawes's research, it is most common in grassy fields and meadows, moist riparian areas, forests of WESTERN REDCEDAR, WESTERN HEMLOCK and

Habitat in Somenos Marsh on Vancouver Island.

DOUGLAS-FIR and disturbed areas. In the southern Interior of BC, VAGRANT SHREW also occurs in dry DOUGLAS-FIR or PONDEROSA PINE forests and steppe grassland. The highest population densities occur in grassy habitats; populations are low in closed forests. Among forested habitats, it seems to prefer WESTERN REDCEDAR with rich moist soils and avoid WESTERN HEMLOCK forests with acidic soils. In southern BC, VAGRANT SHREW becomes scarce above 400 metres elevation. Nevertheless, there are records from higher elevations: 1,450 metres in the Cascade Mountains (Mount Lihumption, now Liumchen Mountain), 1,830 metres in the Purcell Mountains (Mount Revelstoke), 2,133 metres in the Selkirk Mountains (Old Glory Mountain), and 1,652 metres in the Rocky Mountains (Sage Creek).

CINEREUS, DUSKY and PACIFIC SHREWS coexist with VAGRANT SHREWS throughout much of southern BC. Although they live in the same habitats, they exhibit some differences in habitat requirements. In coastal areas, the open grassy habitats preferred by the VAGRANT SHREW are avoided by the PACIFIC SHREW, which is primarily a forest species in BC. In the BC Interior CINEREUS SHREW is often the most common shrew in forested habitats. However, VAGRANT SHREW is the dominant shrew in dry grasslands. VAGRANT SHREW, for example, was the only shrew captured by Geoff Scudder's arthropod traps set in 1997 and 1998 in the Hayne's Lease Ecological Reserve near Osoyoos Lake, a site dominated by ANTELOPE-BUSH.

In her forest study plot in southwestern BC, Myrnal Hawes estimated peak population densities of VAGRANT SHREWS in late summer to be about 12 per hectare. Estimates for a population living in an old uncultivated field in western Washington were much higher, with densities reaching 50 per hectare in late summer. Home range varies with habitat and season. In forested habitats of southwestern BC, the average home range was estimated to be 1,039 square metres for non-breeding animals and 3,258 square metres for breeding animals. However, in the preferred grassland habitat, home ranges are smaller, ranging from 27 to 678 square metres. During the breeding season, males have larger home ranges than females and there is considerable overlap, thus increasing the chances for the sexes to come into contact. In late summer, VAGRANT SHREWS establish territories; individuals have discrete home ranges that do not overlap with those of their neighbours. This may ensure an adequate food supply during winter by reducing competition for food.

During summer and autumn, VAGRANT SHREW is active mostly at night. In spring it appears to have three peaks of activity: in the early evening just after dark, at dawn and in the late afternoon. In captivity, this shrew was active throughout the day and night, but the activity periods lasted only four minutes. Captive animals produced high-frequency sounds (18 to 60 kilohertz) for echolocation.

VAGRANT SHREW constructs several types of nests. Resting nests, made from grass, consist of an open cup 6 to 24 centimetres wide and 4 to 6 centimetres high. Brood nests for rearing young (9 to 14 centimetres in diameter and 5 to 7 centimetres high) have a loose outer layer and a compact inner layer 2 to 3 centimetres in diameter that is lined with fine grass or leaves. Larger nests are constructed in cold weather.

The diet is diverse. In coastal Oregon, about 30 prey types were identified, the major items being insect larvae, spiders, snails, slugs, adult and larval beetles, flies and underground fungi. A population inhabiting montane meadows in eastern Oregon consumed mostly earthworms, caterpillars, spiders, crickets, larval and adult June beetles, slugs and snails. Earthworms and non-flying insects were the predominant prey in grasslands lightly grazed by cattle; in heavily grazed grasslands, flying insects and caterpillars formed the bulk of the shrew's diet. Captive VAGRANT SHREWS cached and ate seeds of various coniferous trees and consumed earthworms, centipedes, small slugs, termites and the carcasses of dead mice.

In the rich low-elevation grassland habitats studied by Myrnal Hawes, males became sexually active in mid-February and females in March just after completing their moult, and the first young appeared in April. The litter size is from two to eight, with five to six most common. The gestation period is about 20 days. Females may produce three litters in the breeding season. First-year breeding is common among females, with as many as 50 per cent breeding in the summer of their birth. Males, however, do not breed in their first summer. In the forested habitats (335 metres elevation) studied by Hawes, the breeding season of the VAGRANT SHREW was shorter and females did not breed in their first summer.

At birth, the naked and blind young weigh about 0.4 grams. They develop rapidly, and by 25 days are weaned and leave the nest. The maximum life span is about 17 months; about 17 per cent of the young will survive from birth to the breeding season the following spring. Predators include snakes and owls, such as the BARN OWL, GREAT HORNED OWL and LONG-EARED OWL.

CONSERVATION STATUS

With a large range extent and broad habitat associations, VAGRANT SHREW is considered secure in BC.

SYSTEMATICS AND TAXONOMY

VAGRANT SHREW is a member of the *Sorex vagrans* group, a complex of North American shrews that also includes the water shrew species. (See page 6 for a list of other species.) Based on molecular genetic studies, the closest relative to VAGRANT SHREW is the ORNATE SHREW. These two species form a distinct clade within the *Sorex vagrans* group.

Four subspecies are recognized; one occurs in BC. Vancouver Island and the southern Gulf Islands populations were originally classified as a subspecies *Sorex vagrans vancouverensis* distinct from mainland populations. Although Neal Woodman listed this subspecies in his monograph, Sarah George and James Smith found considerable overlap in cranial size among Vancouver Island and mainland populations and treated them all as a single subspecies. Populations from the southern Gulf Islands and San Juan Islands of Washington have larger skulls than either the mainland or Vancouver Island populations, but the differences were considered insufficient to warrant the recognition of a separate subspecies. A genetic study is needed to assess the genetic divergence and taxonomy of these island populations.

- *Sorex vagrans vagrans* Baird. A widespread race found throughout much of the western United States, southwestern Alberta and southern BC.

REMARKS

The insular distributions of PACIFIC SHREW and VAGRANT SHREW are curious. They are the only shrews found on the smaller islands off the southern coast of BC. Both inhabit Vancouver Island, the largest island on the BC coast, but they do not co-occur on the smaller adjacent islands. VAGRANT SHREW inhabits the Gulf Islands and San Juan Islands off the southeast coast of Vancouver Island, but on islands to the north and islands off the west coast of Vancouver Island, it is replaced by PACIFIC SHREW. The two species may be too similar in their ecology to coexist on small islands.

The species name *vagrans* is derived from the Latin verb *vagor*, meaning "to wander." It is not clear why the term *wandering* was attached to this species. Its home-range size and daily movements fall within the expected range of other shrews of similar body size.

REFERENCES

Buchler (1976); Demboski and Cook (2001); Eisenberg (1964); George and Smith (1991); Gillihan and Foresman (2004); Hawes (1975, 1977); Hooven, Hoyer and Storm (1975); Huggard and Klenner (1998); Newman (1976); Terry (1978); Whitaker, Cross and Maser (1983); Whitaker and Maser (1976); Woodman (2018).

Hypothetical Species

The following two species are shrews that could be found in northern BC. Currently there are no documented captures to confirm their presence there, but they are species that should be considered in faunal surveys or small-mammal ecological research. Any putative BC occurrences will require voucher specimens and/or genetic samples to confirm their identification.

Potential PRAIRIE SHREW habitat on the Peace River east of Taylor, BC.

Prairie Shrew *Sorex haydeni*

DESCRIPTION AND IDENTIFICATION

The PRAIRIE SHREW has a brown dorsal pelage with greyish undersides. The tail is not bicoloured and lacks a dark tuft at its ventral tip.

Cranial/dental traits: 32 teeth; incisors 1/1, unicuspids 5/1, premolars 1/1, molars 3/3; upper incisor with a large medial tine situated below the upper edge of pigment; third upper unicuspid usually larger than the fourth; a pigmented ridge on the upper unicuspids extends to the cingulum and is not separated by a longitudinal groove; lower incisor has three denticles with continuous pigmentation on the labial side usually extending to second denticle.

Although not closely related, CINEREUS SHREW is the only species in the Peace River region that closely resembles PRAIRIE SHREW. Distinguishing them from external features is difficult. PRAIRIE SHREW differs by a smaller body size, paler pelage and tail, and a pale rather than dark tuft at the ventral tip of its tail. Confirmation will require examining cranial/dental traits or taking a genetic sample. Cranial measurements of PRAIRIE SHREW are smaller than those of CINEREUS SHREW, but show some overlap. However, the length of the upper unicuspid toothrow, which is less than 2.2 millimetres, is diagnostic for PRAIRIE SHREW.

MEASUREMENTS

Measurements are based on 27 specimens from Alberta.

	Mean	Range	Sample size
Total length:	83 mm	68–94 mm	$n = 27$
Tail vertebrae:	31 mm	25–38 mm	$n = 22$
Hind foot:	10 mm	9–11 mm	$n = 27$
Mass:	3.7 g	2.1–7.6 g	$n = 26$

DISTRIBUTION

Associated with the Great Plains, PRAIRIE SHREW is found in Canada in the aspen parklands and grasslands of Alberta, Saskatchewan and Manitoba. A historical museum specimen taken in 1947 at Beaverlodge west of Grand Prairie is an outlying occurrence in Alberta. The location is only 37 kilometres east of the BC border, raising the possibility of its presence in the Peace River region. Look for this shrew in arid habitats on the south-facing slopes along the north shore of the Peace River, particularly from Taylor east to the Alberta border.

REFERENCES

Engley and Norton (2001); Van Zyll de Jong (1980, 1983a).

Potential HOLARCTIC LEAST SHREW habitat by Chilkat Pass at Haines Highway in BC.

Holarctic Least Shrew

Sorex minutissimus
Other common names:
Alaska Tiny Shrew, Eurasian Least Shrew

DESCRIPTION AND IDENTIFICATION

The most striking feature of this shrew is its minute size with a body mass less than 2.5 grams and a total length less than 80 millimetres. The dorsal pelage is brown with greyish sides and light-grey undersides giving it a tricoloured appearance. The tail is bicoloured with a brown upper surface and a nearly whitish pale undersurface.

Cranial/dental traits: skull length less than 14.0 millimetres; 32 teeth; incisors 1/1, unicuspids 5/1, premolars 1/1, molars 3/3; pigment on upper incisor very dark with medial tine situated well below the upper edge of pigment; third upper unicuspid usually larger than the fourth; a pigmented ridge on the upper unicuspids does not extend to the cingulum; lower incisor has three denticles with continuous pigmentation on the labial side extending to the third denticle.

The only species that could be confused with this species is WESTERN PYGMY SHREW. HOLARCTIC LEAST SHREW differs externally by a smaller body size and a slightly tricoloured pelage with a bicoloured tail. WESTERN PYGMY SHREW has a larger skull (skull length greater than 14.0 millimetres), small disc-like third and fifth upper unicuspids, and an upper incisor with a large, long medial tine.

MEASUREMENTS

Measurements are taken from Cook et al. (2016).

	Mean	Range	Sample size
Total length:	70 mm	59–74 mm	*n* = 48
Tail vertebrae:	24 mm	19–27 mm	*n* = 48
Hind foot:	9 mm	8–10 mm	*n* = 48
Mass:	1.8 g	1.0–2.5 g	*n* = 48

DISTRIBUTION

HOLARCTIC LEAST SHREW is widely distributed in northern Europe and Asia. Except for a specimen captured in 2014 from northern Yukon near the Dempster Highway, all known North American occurrences are from Alaska where it was discovered in 1997. This species is associated with tundra, shrub tundra, spruce forest and riparian habitats. Although the nearest occurrences in southeastern Alaska are about 300 kilometres from BC, an isolated population could occur by the Haines Highway similar to the isolated TUNDRA SHREW population found there (see TUNDRA SHREW species account on page 177), a shrew with a similar North American range. Look for HOLARCTIC LEAST SHREW in shrub tundra or riparian habitats near Chilkat Pass and Kelsall Lake, the same locations where TUNDRA SHREW was found.

REFERENCES

Cook et al. (2016); Dokuchaev (1997).

Acknowledgements

Research for this book was made possible by the Royal BC Museum. We are indebted to the museum for giving us access to the research building and its mammal collections during the challenging times of closures for the COVID-19 pandemic and the packing of collections for a move to a new building. We especially thank Gavin Hanke for continual support and encouragement and collection managers Lesley Kennes and Anna Chinn for help with our collection needs and specimen loans. For access to the preparatory lab and assistance finding frozen specimens, we thank Darren Copley. Rob Cannings provided guidance on possible sources of invertebrate images; Henry Choong helped troubleshoot our problems and frustrations with the microscope camera. The museum publications team (Eve Rickert, Eva van Emden, Grace Yaginuma and Jeff Werner) guided us through bringing the book to fruition.

Jacqueline Clare and Robyn Renton, at the British Columbia Conservation Data Centre, prepared the species range maps.

There are too many to list here, but many individuals have contributed specimens or observations in the past that were used somewhere in this handbook. We acknowledge citizen scientists who posted their BC mole observations on iNaturalist that contributed to our range maps. Jim Salt generously shared his database of measurements for Alberta shrew species.

Michael Hames allowed use of all his pencil drawings that appeared in the 1996 edition of this handbook, *Opossums, Shrews and Moles of British Columbia*. Donald Gunn granted permission to use two shrew drawings that were previously published in *An Identification Manual to the Small Mammals of British Columbia* and an unpublished shrew drawing.

Kenneth Catania and Andrew Hendry provided the striking cover images. Other contributors of photographs that appear in the book were John Acorn, Bob Brett, Rob Cannings, Syd Cannings, Darren Copley, Vanessa Craig, Jared Hobbs, Andrew Hope, Bengul Kurtar, Markus Merkens, Daphne Nagorsen, Caroline Penn, Jordan Ryckman, Michael Schmidt (Michael Schmidt Photography Vancouver), Andy Teucher, Douglas Watkinson and David Wong.

We thank Chris Stinson (Beaty Biodiversity Museum) and Bob Timm (University of Kansas) for approving and expediting loans of specimens from their collections.

The following institutions provided data for specimens housed in their collections: Academy of Natural Sciences of Drexel University, Philadelphia; American Museum of Natural History, New York City; Beaty Biodiversity Museum, University of British Columbia, Vancouver; Biodiversity Research and Teaching Collections, Texas A&M University, College Station; Burke Museum, University of Washington, Seattle; California Academy of Sciences,

San Francisco; Canadian Museum of Nature, Ottawa; Field Museum of Natural History, Chicago; KU Natural History Museum, University of Kansas, Lawrence; the Manitoba Museum, Winnipeg; Museum of Comparative Zoology, Harvard University; Museum of the North, University of Alaska, Fairbanks; Museum of Southwestern Biology, University of New Mexico, Albuquerque; Museum of Vertebrate Zoology, University of California, Berkeley; Museum of Zoology, University of Michigan, Ann Arbor; National Museum of Natural History (Smithsonian Institution), Washington, DC; Royal Ontario Museum, Toronto; Slater Museum of Natural History, University of Puget Sound, Tacoma; University of Alberta, Edmonton; and University of Montana, Missoula. The Smithsonian Institution Archives provided photocopies of historical field notes from the United States Biological Survey.

Appendix 1: Other Species Referred to in the Book

Common names used in book	Scientific name
AMERICAN BEAVER	*Castor canadensis*
AMERICAN BULLFROG	*Lithobates catesbeianus*
AMERICAN MARTEN	*Martes americana*
ANTELOPE-BUSH	*Purshia tridentata*
ARCTIC GRAYLING	*Thymallus arcticus*
ARIZONA SHREW	*Sorex arizonae*
ARROWLEAF BALSAMROOT	*Balsamorhiza sagittata*
BARN OWL	*Tyto alba*
BARRED OWL	*Strix varia*
BELTED KINGFISHER	*Megaceryle alcyon*
BIGLEAF MAPLE	*Acer macrophyllum*
BIG SAGEBRUSH	*Artemisia tridentata*
BLACK SPRUCE	*Picea mariana*
BLUE-HEADED VIREO	*Vireo solitarius*
BROOK STICKLEBACK	*Culea inconstans*
CHEATGRASS	*Bromus tectorum*
COASTAL TAILED FROG	*Ascaphus truei*
COMMON SHINER	*Luxilus cornutus*
COMMON SHREW	*Sorex araneus*
DIFFUSE KNAPWEED	*Centaurea diffusa*
DOMESTIC CAT	*Felis catus*
DOMESTIC DOG	*Canis familiaris*
DOUGLAS-FIR	*Pseudotsuga menziesii*
DULL OREGON GRAPE	*Mahonia nervosa*
DWARF BIRCH	*Betula glandulosa*
EASTERN MOLE	*Scalopus aquaticus*
EASTERN PYGMY SHREW	*Sorex hoyi*
EASTERN WATER SHREW	*Sorex albibarbis*
ELEGANT WATER SHREW	*Nectogale elegans*
ENGELMANN SPRUCE	*Picea engelmannii*
ERMINE	*Mustela erminea*
ETRUSCAN SHREW	*Suncus etruscus*
EURASIAN WATER SHREW	*Neomys fodiens*
EUROPEAN MOLE	*Talpa europaea*
FOG SHREW	*Sorex sonomae*
GRAND FIR	*Abies grandis*
GREAT HORNED OWL	*Bubo virginianus*
LABRADOR TEA	*Ledum groenlandicum*
LARCH SAWFLY	*Pristiphora erichsonii*
LODGEPOLE PINE	*Pinus contorta*

Common names used in book	Scientific name
LONG-EARED OWL	*Asio otus*
LONG-TAILED MOLE	*Scaptonyx fusicauda*
MARITIME SHREW	*Sorex maritimensis*
MEADOW VOLE	*Microtus pennsylvanicus*
MEXICAN SPOTTED OWL	*Strix occidentalis lucida*
MONTANE SHREW	*Sorex monticola*
MOUNTAIN SAGEWORT	*Artemisia norvegica*
MT. LYELL SHREW	*Sorex lyelli*
NEEDLE-AND-THREAD GRASS	*Stipa comata*
NORTH AMERICAN OPOSSUM	*Didelphis virginiana*
NORTHERN GRASS-OF-PARNASSUS	*Parnassia palustris*
NORTHERN RACCOON	*Procyon lotor*
NORTHERN SAW-WHET OWL	*Aegolius acadicus*
ORNATE SHREW	*Sorex ornatus*
PACIFIC GIANT SALAMANDER	*Dicamptodon tenebrosus*
PACIFIC MARTEN	*Martes caurina*
PAPER BIRCH	*Betula papyrifera*
PASTURE SAGE	*Artemisia frigida*
PONDEROSA PINE	*Pinus ponderosa*
RABBIT-BRUSH	*Ericameria nauseosus*
RED ALDER	*Alnus rubra*
RED ELDERBERRY	*Sambucus racemosa*
RED FOX	*Vulpes vulpes*
RUBBER BOA	*Charina bottae*
SALMONBERRY	*Rubus spectabilis*
SITKA SPRUCE	*Picea sitchensis*
SKUNK CABBAGE	*Lysichiton americanus*
SNOWBERRY	*Symphoricarpos albus*
SOUTHERN SHORT-TAILED SHREW	*Blarina carolinensis*
SPOTTAIL SHINER	*Notropis hudsonius*
SPRUCE BUDWORM	*Choristoneura fumiferana*
SUBALPINE FIR	*Abies lasiocarpa*
TAMARACK	*Larix laricina*
TIEN SHAN SHREW	*Sorex asper*
TIMBER MILK-VETCH	*Astragalus miser*
TREMBLING ASPEN	*Populus tremuloides*
UNALASKA PAINTBRUSH	*Castilleja unalaschcensis*
WESTERN HEMLOCK	*Tsuga heterophylla*
WESTERN LARCH	*Larix occidentalis*
WESTERN REDCEDAR	*Thuja plicata*
YARROW	*Achillea millefolium*

Appendix 2: External Traits to Identify Shrews

Use the following table with the Identification Key to Live Animals in the Hand and the species accounts. It includes only shrew species currently known to occur in BC.

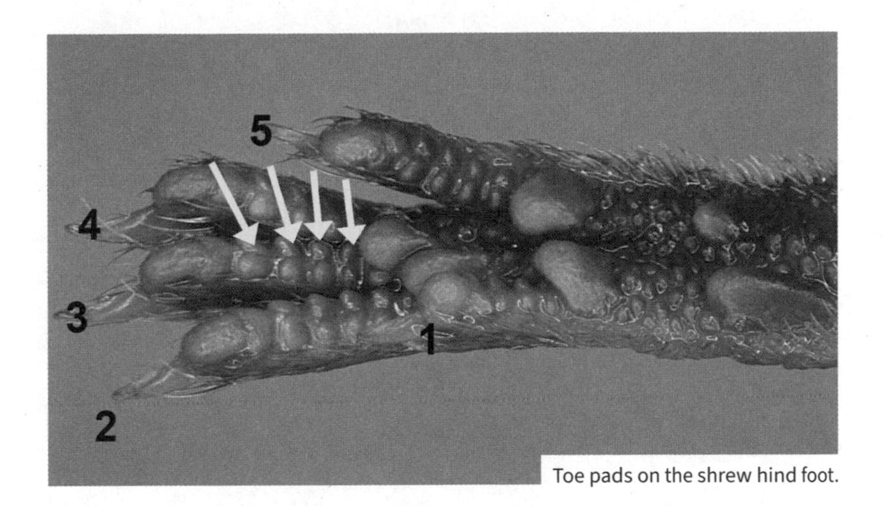

Toe pads on the shrew hind foot.

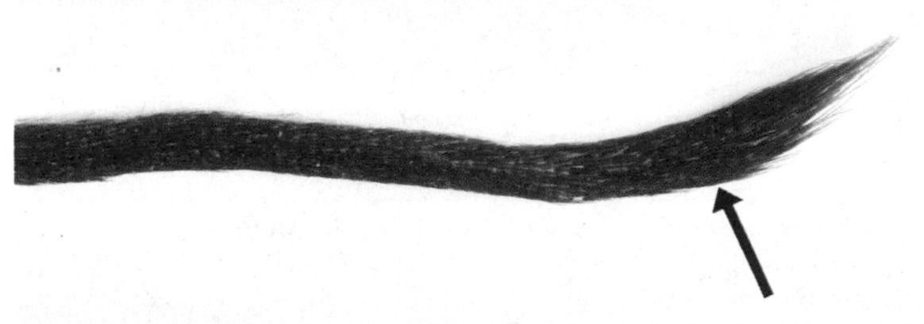

Keel on the distal end of the tail.

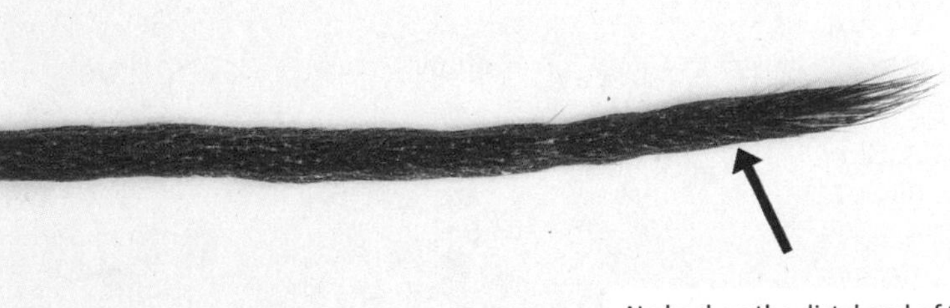

No keel on the distal end of the tail.

Species	Pelage colour	Pelage pattern	Tail colour	Tail mono- or bicoloured	Other diagnostic external traits
ARCTIC SHREW	Back: Dark brown to black; Sides: Paler brown; Underside: Grey to light brown	Tricoloured	Dorsal: Brown to black Ventral: Paler brown	Bicoloured (indistinct)	
PACIFIC WATER SHREW	Back: Dark brown to black; Sides: Paler brown Underside: Paler brown	Monocoloured	Dorsal: Dark brown Ventral: Dark brown	Monocoloured	Feet with stiff fringe of hairs
CINEREUS SHREW	Back: Pale brown (Interior populations), dark brown or grey (coast populations); Underside: Greyish white	Bicoloured	Dorsal: Brown to dark brown Ventral: Pale brown	Bicoloured (indistinct)	3rd to 5th hind toes with 6 pairs of toe pads
WESTERN PYGMY SHREW	Back: Dull greyish brown; Underside: Light grey or brown	Bicoloured	Dorsal: Brown Ventral: Pale brown	Bicoloured (indistinct)	Snout anterior to mouth shorter than other shrew species, tail usually <40% of total length
MERRIAM'S SHREW	Back: Pale greyish brown; Underside: Nearly white	Bicoloured	Dorsal: Brown Ventral: White	Bicoloured (distinct)	
WESTERN WATER SHREW	Back: Dark grey or black; Underside: Silver grey with brown wash	Bicoloured	Dorsal: Dark grey Ventral: White or grey	Bicoloured (distinct)	Feet with stiff fringe of hairs, distal end of tail with a ventral keel
DUSKY SHREW	Back: Brown Sides: Lighter brown; Underside: Brown to grey	Bicoloured	Dorsal: Brown Ventral: Lighter brown	Bicoloured (indistinct)	2nd to 5th hind toes with more than 4 pairs of toe pads

Species	Pelage colour	Pelage pattern	Tail colour	Tail mono- or bicoloured	Other diagnostic external traits
PACIFIC SHREW	Back: Dark brown to near black; Sides: Lighter brown; Underside: Light brown to grey	Bicoloured	Dorsal: Dark brown Ventral: Lighter brown	Bicoloured (indistinct)	2nd to 5th hind toes with more than 4 pairs of toe pads
AMERICAN WATER SHREW	Back: Dark brown or black; Underside: Silver grey sometimes with brownish wash	Bicoloured	Dorsal: Dark brown or black Ventral: White or grey	Bicoloured (distinct)	Feet with stiff fringe of hairs, distal end of tail with a ventral keel
PREBLE'S SHREW	Back: Brown; Underside: Pale brown	Bicoloured	Dorsal: Brown Ventral: Pale brown	Bicoloured (distinct)	3rd and 4th hind toes with 5 pairs of toe pads
OLYMPIC SHREW	Back: Brown; Underside: Pale brown	Bicoloured	Dorsal: Brown Ventral: Pale whitish brown	Bicoloured (distinct)	3rd and 4th hind toes with 5 pairs of toe pads
TROWBRIDGE'S SHREW	Back: Dark or sooty grey; Underside: Lighter grey	Monocoloured	Dorsal: Dark grey Ventral: White	Bicoloured (distinct)	
TUNDRA SHREW	Back: Brown; Sides: Greyish pale brown; Underside: Grey	Tricoloured	Dorsal: Brown Ventral: Buff	Bicoloured (indistinct)	
VAGRANT SHREW	Back: Greyish to dark brown (Interior BC) or darker (coastal BC); Underside: Brown to grey	Bicoloured	Dorsal: Dark brown Ventral: Brown to grey	Bicoloured (indistinct)	2nd to 5th hind toes with 4 pairs or fewer toe pads

Appendix 3: Cranial/Dental Traits to Identify Shrews

Use this table with the Identification Key to Cranial/Dental Traits (page 52) and the species accounts (page 67). It includes only shrew species currently known to occur in BC.

	Upper jaw			Lower jaw			
Species	3rd vs. 4th unicuspid	Medial ridge pigment on 1st unicuspid	Medial tines of incisor	# denticles in labial pigmented area of incisor	Extent of lingual pigment[a]	Postmandibular foramen[b]	Other diagnostic traits
ARCTIC SHREW	3rd taller than 4th	Lower half unpigmented	Small to medium; lower edge well below pigment level	1, 2 or 3	1st denticle	In 100%, on both dentaries; small to large	Hypocone on 1st upper molar pigmented,[c] 1st unicuspid medial ridge unpigmented at medial end, edge of pigment on anterior face of upper incisor sits at or below pigment on lateral cusp
PACIFIC WATER SHREW	3rd equal or shorter than 4th	Weak to strong pigmentation to cingulum	Small to large; lower edge at, below or well below pigment level	3	2nd or 3rd denticle	In 15%, usually on 1 dentary only; tiny to small	Rostrum curved downward in the lateral view
CINEREUS SHREW	3rd equal or taller than 4th	Strongly pigmented to cingulum, ~10% of individuals unpigmented	Small to large; lower edge below or well below pigment level	1, 2 or 3	2nd or 3rd denticle	In 30%, on 1 or both dentaries; small to large	
WESTERN PYGMY SHREW	3rd shorter than 4th	Strongly pigmented ridge does not reach cingulum, but it angles posteriorly as a pigmented ridge	Large; lower edge well below pigment level	3	2nd denticle	Absent	3rd unicuspid tiny and disc-like (from a ventral view), 5th unicuspid hidden behind 1st molar in side view, upper incisors with deep groove between medial tine and incisor

MERRIAM'S SHREW	3rd unicuspid taller than 4th	Strongly pigmented ridge does not reach cingulum	Absent	3	2nd or 3rd denticle	In 100%, on both dentaries; small to large	No medial tine on upper incisors, space between denticles on lower incisor very shallow, 3rd denticle imperceptible
WESTERN WATER SHREW	3rd shorter than 4th	Weak to strong pigmentation to cingulum	Small to medium; lower edge well below pigment level	3	2nd or 3rd denticle	In 50%, mostly on 1 dentary only; tiny to large	
DUSKY SHREW	3rd shorter than 4th	Strongly pigmented to cingulum	Medium; lower edge well below pigment level	3	2nd or 3rd denticle	In 60%, about half of them are on 1 dentary only; tiny to medium	
PACIFIC SHREW	3rd shorter than 4th	Weak to strong pigmentation to cingulum	Medium; lower edge above, at, below or well below pigment level	3	1st, 2nd or 3rd denticle	In 25%, mostly on 1 dentary only; tiny to medium	
AMERICAN WATER SHREW	3rd equal or shorter than 4th	Weak to strong pigmentation to cingulum in ~50% of individuals, ~50% unpigmented	Medium to large; lower edge below or well below pigment level	2 or 3	2nd or 3rd denticle	In 40%, mostly on 1 dentary only; tiny to medium	
PREBLE'S SHREW	3rd equal or taller than 4th	Weak or interrupted pigment to cingulum	Medium; lower edge well below pigment level	2 or 3	2nd denticle	Absent	Unicuspids compressed along toothrow and roughly square in shape

	Upper jaw		Lower jaw				
Species	3rd vs. 4th unicuspid	Medial ridge pigment on 1st unicuspid	Medial tines of incisor	# denticles in labial pigmented area of incisor	Extent of lingual pigment[a]	Postmandibular foramen[b]	Other diagnostic traits
OLYMPIC SHREW	3rd shorter, equal or taller than 4th	Weak pigmentation to cingulum in ~50% of individuals, ~50% unpigmented	Absent to small; lower edge above, at or below pigment level	1	1st or 2nd denticle	In 100%, on both dentaries; small to large	~20% have no medial tine on upper incisors
TROW-BRIDGE'S SHREW	3rd shorter than 4th	Weakly pigmented in ~50% of individuals, ~50% unpigmented	Small to large; lower edge at or below pigment level	2 or 3	2nd denticle	In 100%, on both dentaries; medium to large	1st unicuspid's medial ridge is unpigmented at medial end, separated from cingulum by a groove
TUNDRA SHREW	3rd taller than 4th	Pigmented partial ridge extends about two-thirds of the way to cingulum	Small; lower edge well below pigment level	3	1st denticle	In 100%, on both dentaries; medium to large	Hypocone on 1st upper molar pigmented,[c] 1st unicuspid's medial ridge is unpigmented at medial end, edge of pigment on anterior face of upper incisor sits above pigment of lateral cusp
VAGRANT SHREW	3rd equal or shorter than 4th	Weak to strong pigmentation to cingulum	Medium; lower edge above, at or below pigment level	3	2nd or 3rd denticle	In 45%, mostly on 1 dentary only; tiny to medium	Pigmented medial tine of upper incisor often separated from main pigment area by a pale gap

a See couplet 6 in Identification Key to Cranial/Dental Traits (page 60).

b See couplet 15 in Identification Key to Cranial/Dental Traits (page 65).

c See couplet 6 in Identification Key to Cranial/Dental Traits (page 60).

Glossary

allozymes—Genetic variants of enzymes encoded by structural genes. Before DNA sequencing was developed, allozymes were prominent in genetic studies.

auditory bulla—A bony capsule that covers the middle and inner ear.

biodiversity—Biological diversity, which is the variety of living organisms in an area.

biogeography—The study of the geographical distribution of plants and animals.

Blue List—A list of BC species that are not immediately threatened but are of concern because of characteristics that make them particularly sensitive to human activities or natural events.

canines—The teeth behind the **incisors** and in front of the **premolars**; a single canine occurs on each side of the upper and lower jaws in mammals. Canines are usually long and dagger-like but are flattened in moles. The canines of shrews are not easily distinguished from the incisors and premolars.

cingulum—An enamel shelf that lies below the cusp or peak of a tooth; in shrews the cingulum on the upper **unicuspid** teeth is on the inside or tongue side of the unicuspid teeth.

clade—A group in a **phylogenetic tree** that includes an ancestor and all of its descendants.

community—An assemblage of organisms living together in a particular environment.

coronoid process—The top bony projection on the posterior end of the **dentary**.

COSEWIC—The Committee on the Status of Endangered Wildlife in Canada, a national committee of experts that assess conservation status of Canadian wildlife.

Cretaceous—In the geological time scale, a period in the Mesozoic era that lasted from about 146 to 65.5 million years ago. It is the period when the modern-day **eutherian** mammals first appeared and the dinosaurs were dominant.

cusp—A high peak or rounded area on the crown of a tooth.

Dehnel phenomenon—A shrinkage in body size that occurs in some northern populations of shrews in winter as an adaptation to reduce energy demands.

dentary—The pair of bones that form the **mandible** or lower jaw.

denticle—For shrews, the conical projections or bumps on the top edge of the lower **incisor** tooth.

dentition—The form and arrangement of the teeth.

distal—Refers to sites located away from the centre of the body.

dorsal—On the back or upper side.

echolocation—An orientation system based on generating sounds and listening to their returning echoes to locate obstacles or prey.

ecoprovince—A broad geographic area with consistent climate and terrain; 10 terrestrial ecoprovinces are recognized in British Columbia.

ecosection—Part of an ecological classification system of the province with three levels: ecoprovince, ecoregion and ecosection. Ecosections are the lowest level, defined by minor physiographic and macroclimatic variation; there are 139 of them in BC.

Eimer's organ—Minute touch-sensitive organs on the snout of moles.

endemic—Native to a particular area or region.

Eocene—In the geological time scale, an epoch in the Cenozoic era that lasted from about 56 to 33.9 million years ago.

eutherian—A true placental mammal, one that nourishes its unborn young through a **placenta**. Eutheria is an infraclass in the class Mammalia that includes all living mammals except for marsupials and monotremes.

flank gland—Glands on the sides or flanks of shrews that produce a pungent odour and are associated with breeding. In males, the flank glands appear as rectangular areas, five to eight millimetres long, with tufts of bristles; in females, they are usually only visible on the inside of the skin.

forb—A herbaceous (not woody), broadleaf flowering plant that is not grass-like.

fossorial—Adapted to living underground (e.g., moles).

generalized—Adapted to a wide range of conditions; not specialized.

gestation period—The length of pregnancy; the time from the fertilization of the ovum to the birth of the fetus.

hibernation—A state of lethargy characterized by a reduction in body temperature and **metabolic rate**.

home range—The area covered by an animal during its normal day-to-day activities.

hypocone—A small **cusp** in the corner of an upper molar tooth in mammals.

incisors—The front teeth in the mammalian upper and lower jaws, preceding the **canines**.

IUCN—The International Union for Conservation of Nature, an international organization that is the global authority on the status of the natural world. It produces the IUCN Red List of Threatened Species, a conservation tool.

kilohertz—A unit for measuring the frequency of sound; one kilohertz (kHz) is equal to 1,000 cycles per second.

labial—The side of a tooth facing the lips and cheek.

larva—The immature form of an insect that differs greatly from the adult; for example, caterpillar or grub.

Lepidoptera—An order of insects that includes moths and butterflies.

lineage—A branch of a **phylogenetic tree**; an ancestral species and its descendants.

lingual—The side of a tooth facing the tongue.

mandible—The lower jaw, made up of two **dentary** bones.

maxilla—A paired bone of the skull that bears all the upper teeth except the incisors.

medial—Toward the middle or centre.

metabolic rate—A measure of the general activity level of an animal; free energy production per unit of body mass.

Miocene—In the geological time scale, an epoch in the Cenozoic era that lasted from 23 to 5.3 million years ago.

mitochondrial DNA—A circular strand of DNA found in the mitochondrion of cells, which encodes some of the proteins used in cellular metabolism, generally following a maternal inheritance pattern. Because it has high mutation rates compared with **nuclear DNA**, it is used for establishing relationships among groups that recently diverged.

molars—The teeth located behind the **premolars**; they first appear after the deciduous milk teeth are shed.

molecular clock—A measure of evolutionary change over time at the molecular level that is based on the mutation rate of DNA sequences.

molecular genetics—A field of biology that studies the structure and functions of genes at a molecular level by analyzing DNA. Data from molecular genetics are used to create **phylogenetic trees**.

morphometrics—Measurements that describe the size and shape of a form.

moult—The shedding of hairs and their subsequent replacement with new hair or fur.

museum study skin—An archival research specimen consisting of a dried skin with fur filled with cotton.

nuclear DNA—DNA contained in the nucleus of cells (as chromosomes), accounting for most of the genetic material in the cell. It is typically inherited equally from both parents in contrast to **mitochondrial DNA**.

nymph—The immature form of an insect that usually resembles an adult; in aquatic insects the nymphs have gills.

Oligocene—In the geological time scale, an epoch in the Cenozoic era that lasted from 33.9 to 23 million years ago.

parr—A young salmon or trout.

pelage—The fur or hair covering of a mammal.

phylogenetic tree—An illustration or diagram that depicts the evolutionary relationship or kinship among a group of organisms (e.g., species).

phylogeny—The study of the history of evolution and the lines of descent in a group of organisms.

placenta—A membranous structure formed by fetal and maternal tissues that connects the embryo to the uterus of the mother and facilitates the exchange of nutrients and waste. Only the **eutherian** mammals possess a complete placenta.

Pliocene—In the geological time scale, an epoch in the Cenozoic era that lasted from 5.3 to 2.6 million years ago.

postmandibular foramen—An opening on the **medial** surface of the back of the shrew **mandible** that allows the entrance of a nerve and vessels. It can appear as a separate opening beside the mandibular foramen or it can be confluent within the same large opening of the mandibular foramen.

premaxilla—A paired bone in the front of the skull that bears the upper **incisor** teeth.

premolars—The teeth located behind the **canine** and in front of the **molars** in the mammalian jaws. The number of premolars varies among mammals, but the maximum is four on each side of the upper and lower jaws, as in moles.

pupa—The inactive immature form of an insect; the resting stage where the insect is encased in a cocoon.

Red List—A list of species that have been legally designated as endangered or threatened under the BC Wildlife Act, are extirpated or are candidates for such designation.

relict—A population that persists locally after the extinction of the species.

resorbed—When milk teeth are shed and then absorbed by the body.

rostrum—The nasal area or snout of a skull.

runway—A small worn trail made by small mammals in grassy or meadow habitats.

Species at Risk Act (SARA)—The federal legislation that protects endangered or threatened organisms and their habitats.

species concepts—The various criteria that are used to define a species.

species diversity—The variety of species in an area, often measured as the number of species. Also called species richness.

subnivean—An opening beneath the snow where small mammals live during winter.

subspecies—Populations of a species that are geographically separated and differ taxonomically from other populations; sometimes called geographic races.

suture—Point of contact or juncture between adjacent bones.

tine—In shrews the small, pointed accessory **cusp** in the middle of the anterior face of the upper **incisor**. The presence or absence of these **medial** tines and their size and position relative to the pigmented area on the face of the incisor are an important identification trait.

torpor—A daily short-term state of inactivity achieved by lowering the body temperature and reducing the metabolic rate in order to conserve energy.

trap nights—An index of trapping effort in small-mammal studies, which is based on the number of traps used multiplied by the number of trapping nights.

type locality—The locality where the **type specimen** was collected.

type specimen—The single specimen designated by the original author as the type associated with the author's description of the species or **subspecies**.

unicuspid—A tooth with one **cusp**. All BC shrew species have a series of five unicuspids in the upper jaw and one in the lower behind the **incisors**. The relative size of the upper unicuspids is an important identification trait.

ventral—On the underside or bottom.

ventral keel—In the AMERICAN WATER SHREW and WESTERN WATER SHREW, the ventral surface of the tail for about the last centimetre from the tip has a raised fleshy ridge about one millimetre high fringed by stiff hairs. This forms a widening of the tail compared with other shrews and presumably functions to aid propulsion in water.

vibrissae—Long tactile hairs on the face; whiskers.

voucher—A preserved specimen, photograph or genetic sample that serves as a verifiable and permanent record of a plant or animal species.

weaning—When mammalian young finish nursing and begin to eat solid food.

Yellow List—A list of BC species that are apparently secure and not at risk of extinction.

young-of-the-year—A small mammal in its first year of life.

zygomatic arch—An arch of bone that extends across the orbit of the eye in the skull. Not found in the shrew skull but present in mole skulls.

References

Aitchison, C.W. 1987a. "Review of Winter Trophic Relations of Soricine Shrews." *Mammal Review* 17, no. 1: 1–24.

———. 1987b. "Winter Energy Requirements of Soricine Shrews." *Mammal Review* 17, no. 1: 25–38.

Alexander, L.F. 1996. *A Morphometric Analysis of Geographic Variation within the* Sorex monticolus *(Insectivora: Soricidae)*. Miscellaneous Publication 88. Lawrence, KS: University of Kansas, Museum of Natural History.

Amori, G., F. Chiozza, C. Rondinini and L. Luiselli. 2011. "Worldwide Conservation Hotspots for Soricomorpha Focusing on Endemic Island Taxa: An Analysis at Two Taxonomic Levels." *Endangered Species Research* 15, no. 2: 143–149.

Anderson, R.M. 1934. "Sorex palustris brooksi, a New Water Shrew from Vancouver Island." *Canadian Field-Naturalist* 48, no. 8: 134.

Anthony, N.M., C.A. Ribic, R. Bautz and T. Garland Jr. 2005. "Comparative Effectiveness of Longworth and Sherman Live Traps." *Wildlife Society Bulletin* 33, no. 3: 1018–1026.

Armstrong, D.M., and J.K. Jones Jr. 1971. "Sorex merriami." *Mammalian Species*, no. 2: 1–2.

Armstrong, F.H. 1957. "Notes on *Sorex preblei* in Washington State." *Murrelet* 38, no. 1: 6.

Aubry, K.B., M.J. Crites and S.D. West. 1991. "Regional Patterns of Small Mammal Abundance and Community Composition in Oregon and Washington." In *Wildlife and Vegetation of Unmanaged Douglas-fir Forests*, edited by L.F. Ruggiero, K.B. Aubry, A.B. Carey and M.H. Huff, 285–294. General Technical Report PNW-GTR-285. Portland, OR: US Department of Agriculture, Forest Service.

Baird, D.D., R.M. Timm and G.E. Nordquist. 1983. "Reproduction in the Arctic Shrew, *Sorex arcticus*." *Journal of Mammalogy* 64, no. 2: 298–301.

Baker, R.J., and R.D. Bradley. 2006. "Speciation in Mammals and the Genetic Species Concept." *Journal of Mammalogy* 87, no. 4: 643–662.

BC Ministry of Environment. 2014. *Recovery Plan for the Townsend's Mole (*Scapanus townsendii*) in British Columbia*. Victoria, BC: BC Ministry of Environment.

Bee, J.W., and E.R. Hall. 1956. *Mammals of Northern Alaska on the Arctic Slope*. Miscellaneous Publication 8. Lawrence, KS: University of Kansas, Museum of Natural History. https://doi.org/10.5962/bhl.title.63916.

Beneski, J.T., and D.W. Stinson. 1987. "Sorex palustris." *Mammalian Species*, no. 296: 1–6.

Boyle, C.A., L. Lavkulich, H. Schreier and E. Kiss. 1997. "Changes in Land Cover and Subsequent Effects on Lower Fraser Basin Ecosystems from 1827 to 1990." *Environmental Management* 21, no. 2: 185–196.

Branis, M., and H. Burda. 1994. "Visual and Hearing Biology of Shrews." In *Advances in the Biology of Shrews*, edited by J.F. Merritt, G.L. Kirkland Jr. and R.K. Rose, 189–200. Pittsburgh, PA: Carnegie Museum of Natural History.

Brooks, A. 1902. "Mammals of the Chilliwack District." *Ottawa Naturalist* 15, no. 11: 239–244.

Buchler, E.R. 1976. "The Use of Echolocation by the Wandering Shrew (*Sorex vagrans*)." *Animal Behaviour* 24, no. 4: 858–873.

Buckner, C.H. 1957. "Population Studies on Small Mammals of Southeastern Manitoba." *Journal of Mammalogy* 38, no. 1: 87–97.

———. 1964. "Metabolism, Food Capacity and Feeding Behavior in Four Species of Shrews." *Canadian Journal of Zoology* 42, no. 2: 259–279.

———. 1966. "Populations and Ecological Relationships of Shrews in Tamarack Bogs of Southeastern Manitoba." *Journal of Mammalogy* 47, no. 2: 181–194.

———. 1970. "Direct Observation of Shrew Predation on Insects and Fish." *Blue Jay* 28, no. 4: 171–172.

Buckner, C.H., and D.G.H. Ray. 1968. "Notes on the Water Shrew in Bog Habitats of Southeastern Manitoba." *Blue Jay* 26, no. 2: 95–96.

Burgin, C.J., J.P. Colella, P.L. Kahn and N.S. Upham. 2018. "How Many Species of Mammals Are There?" *Journal of Mammalogy* 99, no. 1: 1–14.

Burles, D.W., A.G. Edie and P.M. Bartier. 2004. *Native Land Mammals and Amphibian of Haida Gwaii with Management Implications for Gwaii Haanas National Park Reserve and Haida Heritage Site*. Ottawa, ON: Parks Canada.

Calder, W.A. 1969. "Temperature Relations and Underwater Endurance of the Smallest Homeothermic Diver, the Water Shrew." *Comparative Biochemistry and Physiology* 30, no. 6: 1075–1082.

Campbell, K.L., and P.W. Hochachka. 2000. "Thermal Biology and Metabolism of the American Shrew-mole, *Neurotrichus gibbsii*." *Journal of Mammalogy* 81, no. 2: 578–585.

Campbell, R.W., D.A. Manuwal and A.S. Harestad. 1987. "Food Habits of the Common Barn-Owl in British Columbia." *Canadian Journal of Zoology* 65, no. 3: 578–586.

Carey, A.B., and C.A. Harrington. 2001. "Small Mammals in Young Forests: Implications for Management for Sustainability." *Forest Ecology and Management* 154, no. 1–2: 289–309.

Carey, A.B., and M.L. Johnson. 1995. "Small Mammals in Managed, Naturally Young, and Old-Growth Forests." *Ecological Applications* 5, no. 2: 336–352.

Carraway, L.N. 1987. "Analysis of Characters for Distinguishing *Sorex trowbridgii* from Sympatric *S. Vagrans*." *Murrelet* 68, no. 1: 29–30.

———. 1995. *A Key to Recent Soricidae of the Western United States and Canada Based Primarily on Dentaries*. Occasional Papers of the Museum of Natural History 175. Lawrence, KS: University of Kansas.

Carraway, L.N., L.F. Alexander and B.J. Verts. 1993. "Scapanus townsendii." *Mammalian Species*, no. 434: 1–7.

Carraway, L.N., and B.J. Verts. 1991. "Neurotrichus gibbsii." *Mammalian Species*, no. 387: 1–7.

———. 1994. "Relationship of Mandibular Morphology to Relative Bite Force in Some *Sorex* from Western North America." In *Advances in the Biology of Shrews*, edited by J.F. Merritt, G.L. Kirkland Jr. and R.K. Rose, 201–210. Pittsburgh, PA: Carnegie Museum of Natural History.

———. 1999. "Records of Reproduction in *Sorex preblei*." *Northwestern Naturalist* 80, no. 3: 115–116.

———. 2005. "Assessment of Variation in Cranial and Mandibular Dimensions in Geographic Races of *Sorex trowbridgii*." In *Advances in the Biology of Shrews II*, edited by J.F. Merritt, S. Churchfield, R. Hutter and B.I. Sheftel, 139–153. New York, NY: International Society of Shrew Biologists.

Catania, K.C. 2013. "The Neurobiology and Behavior of the American Water Shrew (*Sorex palustris*)." *Journal of Comparative Physiology* 199, no. 6: 545–554.

Catania, K.C., J.F. Hare and K.L. Campbell. 2008. "Water Shrews Detect Movement, Shape, and Smell to Find Prey Underwater." *PNAS* 105, no. 2: 571–576.

Churchfield, S. 1990. *The Natural History of Shrews*. Ithaca, NY: Comstock.

———. 1994. "Foraging Strategies of Shrews, and the Evidence from Field Studies." In *Advances in the Biology of Shrews*, edited by J.F. Merritt, G.L. Kirkland Jr. and R.K. Rose, 77–87. Pittsburgh, PA: Carnegie Museum of Natural History.

Churchfield, S., J. Barber and C. Quinn. 2000. "A New Survey Method for Water Shrews (*Neomys fodiens*) Using Baited Tubes." *Mammal Review* 30, no. 3–4: 249–254.

Churchfield, S., L. Rychlik and J.R.E. Taylor. 2012. "Food Resources and Foraging Habits of the Common Shrew, *Sorex araneus*: Does Winter Food Shortage Explain Dehnel's Phenomenon?" *Oikos* 121, no. 10: 1593–1602.

Clough, G.C. 1963. "Biology of the Arctic Shrew, Sorex arcticus." *American Midland Naturalist* 69, no. 1: 69–81.

Conaway, C.H. 1952. "Life History of the Water Shrew (*Sorex palustris navigator*)." *American Midland Naturalist* 48, no. 1: 219–248.

Cook, J.A., N.G. Dawson and S.O. MacDonald. 2006. "Conservation of Highly Fragmented Systems: The North Temperate Alexander Archipelago." *Biological Conservation* 133, no. 1: 1–15.

Cook, J.A., B.S. McLean, D.J. Jackson, J.P. Colella, S.E. Greiman, V.V. Tkach, T.S. Jung and J.L. Dunnum. 2016. "First Record of the Holarctic Least Shrew (*Sorex minutissimus*) and Associated Helminths from Canada: New Light on Northern Pleistocene Refugia." *Canadian Journal of Zoology* 94, no. 5: 367–372.

Cornely, J.E., L.N. Carraway and B.J. Verts. 1992. "Sorex preblei." *Mammalian Species*, no. 416: 1–3.

COSEWIC. 2003. *COSEWIC Assessment and Update Status Report on the Townsend's Mole* Scapanus townsendii *in Canada*. Ottawa, ON: Committee on the Status of Endangered Wildlife in Canada.

———. 2006. *COSEWIC Assessment and Update Status Report on the Pacific Water Shrew* Sorex bendirii *in Canada*. Ottawa, ON: Committee on the Status of Endangered Wildlife in Canada.

Craig, V.J. 1995. "Relationships between Shrews (*Sorex* spp.) and Downed Wood in the Vancouver Watersheds." MSc thesis, University of British Columbia, Faculty of Forestry.

———. 2004. *The Status of the Vancouver Island Water Shrew (*Sorex palustris brooksi*) in British Columbia*. Wildlife Bulletin no. B. 114. Victoria, BC: BC Ministry of Sustainable Resource Management.

Dalquest, W.W., and D.R. Orcutt. 1942. "The Biology of the Least Shrew-mole, Neurotrichus gibbsii minor." *American Midland Naturalist* 27, no. 2: 387–401.

Demarchi, D.A. 2011. *An Introduction to the Ecoregions of British Columbia*. Victoria, BC: BC Ministry of Environment.

Demboski, J.R., and J.A. Cook. 2001. "Phylogeography of the Dusky Shrew, *Sorex monticolus* (Insectivora, Soricidae): Insight into Deep and Shallow History in Northwestern North America." *Molecular Ecology* 10, no. 5: 1227–1240.

Diersing, V.E. 1980. "Systematics and Evolution of the Pygmy Shrews (Subgenus *Microsorex*) of North America." *Journal of Mammalogy* 61, no. 1: 76–101.

Diersing, V.E., and D.F. Hoffmeister. 1977. "Revision of the Shrews *Sorex merriami* and a Description of a New Species of the Subgenus *Sorex*." *Journal of Mammalogy* 58, no. 3: 321–333.

Dinets, V. 2017. "Surface Foraging in *Scapanus* Moles." *Mammalia* 82, no. 1: 48–53.

Dokuchaev, N.E. 1997. "A New Species of Shrew (Soricidae, Insectivora) from Alaska." *Journal of Mammalogy* 78, no. 3: 811–817.

Douady, C.J., and E.J.P. Douzery. 2009. "Hedgehogs, Shrews, Moles, and Solenodons (Eulipotyphla)." In *The Timetree of Life*, edited by S.B. Hedges and S. Kumar, 495–498. Oxford, UK: Oxford University Press.

Doyle, A.T. 1990. "Use of Riparian and Upland Habitats by Small Mammals." *Journal of Mammalogy* 71, no. 1: 14–23.

Dubey, S., N. Salamin, S.D. Ohdachi, P. Barrière and P. Vogel. 2007. "Molecular Phylogenetics of Shrews (Mammalia: Soricidae) Reveal Timing of Transcontinental Colonizations." *Molecular Phylogenetics and Evolution* 44, no. 1: 126–137.

Eisenberg, J.F. 1964. "Studies on the Behavior of Sorex vagrans." *American Midland Naturalist* 72, no. 2: 417–425.

Engley, L., and M. Norton. 2001. *Distribution of Selected Small Mammals in Alberta*. Alberta Species at Risk Report 12. Edmonton, AB: Alberta Sustainable Resource Development.

Environment Canada. 2014. *Recovery Strategy for the Pacific Water Shrew (*Sorex bendirii*) in Canada. Species at Risk Act* Recovery Strategy Series. Ottawa, ON: Environment Canada.

———. 2016. *Recovery Strategy for the Townsend's Mole (*Scapanus townsendii*) in Canada. Species at Risk Act* Recovery Strategy Series. Ottawa, ON: Environment Canada.

Foresman, K.R. 1999. "Distribution of the Pygmy Shrew, *Sorex hoyi*, in Montana and Idaho." *Canadian Field-Naturalist* 113, no. 4: 681–683.

Forsyth, D.J. 1976. "A Field Study of Growth and Development of Nestling Masked Shrews (*Sorex cinereus*)." *Journal of Mammalogy* 57, no. 4: 708–721.

Foster, J.B. 1965. *The Evolution of the Mammals of the Queen Charlotte Islands, British Columbia*. Occasional Papers of the British Columbia Provincial Museum 14. Victoria, BC: British Columbia Provincial Museum.

Fraker, M.A., C. Bianchini and I. Robertson. 1999. *Burns Bog Ecosystem Review: Small Mammals*. For Delta Fraser Properties Partnership and the Environmental Assessment Office. Sidney, BC: Roberston Environmental Services and Terramar Environmental Research.

Gashwiler, J.S. 1976. "Notes on the Reproduction of Trowbridge Shrews in Western Oregon." *Murrelet* 57, no. 3: 58–62.

Genoud, M. 1988. "Energetic Strategies of Shrews: Ecological Constraints and Evolutionary Implications." *Mammal Review* 18, no. 4: 173–193.

George, S.B. 1988. "Systematics, Historical Biogeography, and Evolution of the Genus *Sorex*." *Journal of Mammalogy* 69, no. 3: 443–461.

———. 1989. "Sorex trowbridgii." *Mammalian Species*, no. 337: 1–5.

George, S.B., and J.D. Smith. 1991. "Inter- and Intraspecific Variation among Coastal and Island Populations of *Sorex monticolus* and *Sorex vagrans* in the Pacific Northwest." In *The Biology of the Soricidae*, edited by J.S. Findley and T.L. Yates, 75–91. Albuquerque, NM: University of New Mexico.

Giacometti, L., and H. Machida. 1965. "The Skin of the Mole (*Scapanus townsendii*)." *Anatomical Record* 153, no. 1: 31–39.

Giger, R.D. 1973. "Movements and Homing in Townsend's Mole Near Tillamook, Oregon." *Journal of Mammalogy* 54, no. 3: 648–659.

Gillihan, S.W., and K.R. Foresman. 2004. "Sorex vagrans." *Mammalian Species*, no. 744: 1–5.

Gitzen, R.A., J.E. Bradley, M.R. Kroeger and S.D. West. 2009. "First Record of Preble's Shrew (*Sorex preblei*) in the Northern Columbia Basin, Washington." *Northwestern Naturalist* 90, no. 1: 41–43.

Gitzen, R.A., S.D. West and B.E. Trim. 2001. "Additional Information on the Distributions of Small Mammals at the Hanford Site, Washington." *Northwest Science* 75, no. 4: 350–362.

Glendenning, R. 1959. "Biology and Control of the Coast Mole, *Scapanus orarius orarius* True, in British Columbia." *Canadian Journal of Animal Science* 39, no. 1: 34–44.

Gomez, D.M., and R.G. Anthony. 1998. "Small Mammal Abundance in Riparian and Upland Areas of Five Seral Stages in Western Oregon." *Northwest Science* 72, no. 4: 293–302.

Gorman, M.L., and R.D. Stone. 1990. *The Natural History of Moles*. Ithaca, NY: Comstock.

Guiguet, C.J. 1953. *An Ecological Study of Goose Island, British Columbia, with Special Reference to Mammals and Birds*. Occasional Papers of the British Columbia Provincial Museum 10. Victoria, BC: British Columbia Provincial Museum.

Gunther, P.M., B.S. Horn and G.D. Babb. 1983. "Small Mammal Populations and Food Selection in Relation to Timber Harvest Practices in the Western Cascade Mountains." *Northwest Science* 57, no. 1: 32–44.

Gusztak, R.W. 2008. "Dive Performance and Aquatic Thermoregulation of the World's Smallest Mammalian Diver, the American Water Shrew (*Sorex palustris*)." MS thesis, University of Manitoba, Department of Biological Sciences.

Gusztak, R.W., and K.L. Campbell. 2004. "Growth, Development and Maintenance of American Water Shrews (*Sorex palustris*) in Captivity." *Mammal Study* 29, no. 1: 65–72.

Gusztak, R.W., R.A. MacArthur and K.L. Campbell. 2005. "Bioenergetics and Thermal Physiology of American Water Shrews (*Sorex palustris*)." *Journal of Comparative Physiology B* 175, no. 2: 87–95.

Hanski, I. 1994. "Population Biological Consequences of Body Size in *Sorex*." In *Advances in the Biology of Shrews*, edited by J.F. Merritt, G.L. Kirkland Jr. and R.K. Rose, 15–26. Pittsburgh, PA: Carnegie Museum of Natural History.

Hartman, G.D., and T.L. Yates. 1985. "Scapanus orarius." *Mammalian Species*, no. 253: 1–5.

Hawes, M.L. 1975. "Ecological Adaptations in Two Species of Shrews." PhD diss, University of British Columbia, Department of Zoology.

———. 1976. "Odor as a Possible Isolating Mechanism in Sympatric Species of Shrews (*Sorex vagrans* and *Sorex obscurus*)." *Journal of Mammalogy* 57, no. 2: 404–406.

———. 1977. "Home Range, Territoriality and Ecological Separation in Sympatric Shrews, *Sorex vagrans* and *Sorex obscurus*." *Journal of Mammalogy* 58, no. 3: 354–367.

He, K., A. Shinohara, K.M. Helgen, M.S. Springer, X.-L. Jiang and K.L. Campbell. 2017. "Talpid Mole Phylogeny Unites Shrew Moles and Illuminates Overlooked Cryptic Species Diversity." *Molecular Biology and Evolution* 34, no. 1: 78–87.

Hooven, E.F., R.F. Hoyer and R.M. Storm. 1975. "Notes on the Vagrant Shrew, *Sorex vagrans*, in the Willamette Valley of Western Oregon." *Northwest Science* 49, no. 3: 163–173.

Hope, A.G., N. Panter, J.A. Cook, S.L. Talbot and D.W. Nagorsen. 2014. "Multilocus Phylogeography and Systematic Revision of North American Water Shrews (Genus: *Sorex*)." *Journal of Mammalogy* 95, no. 4: 722–738.

Hope, A.G., K.A. Speer, J.R. Demboski, S.L. Talbot and J.A. Cook. 2012. "A Climate for Speciation: Rapid Spatial Diversification within the *Sorex cinereus* Complex of Shrews." *Molecular Phylogenetics and Evolution* 64, no. 3: 671–684.

Hope, A.G., R.B. Stephens, S.D. Mueller, V.V. Tkach and J.R. Demboski. 2020. "Speciation of North American Pygmy Shrews (Eulipotyphla: Soricidae) Supports Spatial but Not Temporal Congruence of Diversification among Boreal Species." *Biological Journal of the Linnean Society* 129, no. 1: 41–60.

Hope, A.G., E. Waltari, V.B. Fedorov, A.V. Goropashnaya, S.L. Talbot and J.A. Cook. 2011. "Persistence and Diversification of the Holarctic Shrew, *Sorex tundrensis* (Family Soricidae), in Response to Climate Change." *Molecular Ecology* 20, no. 20: 4346–4370.

Horvath, O. 1965. "Arboreal Predation on Bird's Nest by Masked Shrew." *Journal of Mammalogy* 46, no. 3: 495.

Hudson, G.E., and M. Bacon. 1956. "New Records of *Sorex merriami* for Eastern Washington." *Journal of Mammalogy* 37, no. 3: 436–438.

Huggard, D., and W. Klenner. 1998. "Effects of Harvest Type and Edges on Shrews at the Opax Mountain Silvicultural Systems Site." In *Managing the Dry Douglas-fir Forests of the Southern Interior*, edited by A. Vyse, C. Hollstedt and D. Huggard, 235–245. Workshop proceedings, April 29–39, 1997, paper 34/1998. Kamloops, BC: Research Branch, BC Ministry of Forests.

Hutterer, R. 1985. "Anatomical Adaptations of Shrews." *Mammal Review* 15, no. 1: 43–55.

———. 2005. "Homology of Unicuspids and Tooth Nomenclature in Shrews." In *Advances in the Biology of Shrews II*, edited by J.F. Merritt, S. Churchfield, R. Hutter and B.I. Sheftel, 397–403. New York, NY: International Society of Shrew Biologists.

Hyvärinen, H. 1994. "Brown Fat and the Wintering of Shrews." In *Advances in the Biology of Shrews*, edited by J.F. Merritt, G.L. Kirkland Jr. and R.K. Rose, 259–266. Pittsburgh, PA: Carnegie Museum of Natural History.

Ivanter, E.V. 1994. "The Structure and Adaptive Peculiarities of Pelage in Soricine Shrews." In *Advances in the Biology of Shrews*, edited by J.F. Merritt, G.L. Kirkland Jr. and R.K. Rose, 441–454. Pittsburgh, PA: Carnegie Museum of Natural History.

Jackson, M.F. 1951. "Variation in an Isolated Population of Shrews of the *Sorex vagrans-obscurus* Group." MS thesis, University of British Columbia.

Jameson, E.W., Jr. 1955. "Observations on the Biology of *Sorex trowbridgei* in the Sierra Nevada, California." *Journal of Mammalogy* 36, no. 3: 339–345.

Johnson, M.L., and C.W. Clanton. 1954. "Natural History of *Sorex merriami* in Washington State." *Murrelet* 35, no. 1: 1–4.

Jung, T.S. 2016. "Predation of a Western Water Shrew (*Sorex navigator*) by a Belted Kingfisher (*Megaceryle alcyon*)." *Canadian Field-Naturalist* 130, no. 4: 299–301.

Jung, T.S., A. Milani, O.E. Barker and N.P. Millar. 2011. "American Pygmy Shrew, *Sorex hoyi*, Consumed by an Arctic Grayling, *Thymallus arcticus*." *Canadian Field-Naturalist* 125, no. 3: 255–256.

Junge, J.A., and R.S. Hoffmann. 1981. *An Annotated Key to the Long-tailed Shrews (Genus* Sorex*) of the United States and Canada, with Notes on Middle American* Sorex. Occasional Papers of the Museum of Natural History 94. Lawrence, KS: University of Kansas.

Junge, J.A., R.S. Hoffmann and R.W. Debry. 1983. "Relationships within the Holarctic *Sorex arcticus–Sorex tundrensis* Species Complex." *Acta Theriologica* 28, no. 21: 339–350.

Kennerley, R.J., T.E. Lacher Jr., M.A. Hudson, B. Long, S.D. McCay, N.S. Roach, S.T. Turvey and R.P. Young. 2021. "Global Patterns of Extinction Risk and Conservation Needs for Rodentia and Eulipotyphla." *Diversity and Distributions* 27, no. 9: 1792–1806.

Kirkland, G.L., and D.F. Schmidt. 1996. "Sorex arcticus." *Mammalian Species*, no. 524: 1–5.

Kowalski, K., and L. Rychlik. 2021. "Venom Use in Eulipotyphlans: An Evolutionary and Ecological Approach." *Toxins* 13, no. 3: 231.

Kuhn, L.W., W.Q. Wick and R.J. Pedersen. 1966. "Breeding Nests of Townsend's Mole in Oregon." *Journal of Mammalogy* 47, no. 2: 239–249.

Lehmkuhl, J.F., R.D. Peffer and M.A. O'Connell. 2008. "Riparian and Upland Small Mammals on the East Slope of the Cascade Range, Washington." *Northwest Science* 82, no. 2: 94–107.

Lewis, T.H. 1983. "The Anatomy and Histology of the Rudimentary Eye of Neurotrichus." *Northwest Science* 57, no. 1: 8–15.

Lloyd, K.J., and J.J. Eberle. 2008. "A New Talpid from the Late Eocene of North America." *Acta Palaeontologica Polonica* 53, no. 3: 539–543.

Mackiewicz, P., M. Moska, H. Wierzbicki, P. Gagat and D. Mackiewicz. 2017. "Evolutionary History and Phylogeographic Relationships of Shrews from *Sorex araneus* Group." *PLOS ONE* 12, no. 6: e0179760.

Maldonado, J.E., S. Young, L.H. Simons, S. Stone, L.D. Parker and J. Ortega Reyes. 2015. "Conservation Genetics and Phylogeny of the Arizona Shrew in the 'Sky Islands' of the Southwestern United States." *Therya* 6, no. 2: 401–420.

Marasco, P.D., P.R. Tsuruda, D.M. Bautista, D. Julius and K.C. Catania. 2007. "Neuroanatomical Evidence for Segregation of Nerve Fibers Conveying Light Touch and Pain Sensation in Eimer's Organ of the Mole." *PNAS* 103, no. 24: 9339–9344.

Martell, A.M., and A.M. Pearson. 1978. "The Small Mammals of the Mackenzie Delta Region, Northwest Territories, Canada." *Arctic* 31, no. 4: 475–488.

Maser, C., and J.F. Franklin. 1974. *Checklist of Vertebrate Animals of the Cascade Head Experimental Forest.* Portland, OR: US Department of Agriculture, Forest Service.

McCabe, T.T., and I. McTaggart Cowan. 1945. "*Peromyscus maniculatus macrorhinus* and the Problem of Insularity." *Transactions of the Royal Canadian Institute* 25, part 2, no. 54: 117–216.

McComb, W.C., K. McGarigal and R.G. Anthony. 1993. "Small Mammal and Amphibian Abundance in Streamside and Upslope Habitats of Mature Douglas-fir Stands, Western Oregon." *Northwest Science* 67, no. 1: 7–15.

McIntyre, I.W., K.L. Campbell and R.A. MacArthur. 2002. "Body Oxygen Stores, Aerobic Dive Limits and Diving Behaviour of the Star-nosed Mole (*Condylura cristata*) and Comparisons with Non-aquatic Talpids." *Journal of Experimental Biology* 205, no. 1: 45–54.

McKey-Fender, D., W.M. Fender and V.G. Marshall. 1994. "North American Earthworms Native to Vancouver Island and the Olympic Peninsula." *Canadian Journal of Zoology* 72, no. 7: 1325–1339.

McTaggart Cowan, I. 1939. *The Vertebrate Fauna of the Peace River District of British Columbia.* Occasional Papers of the British Columbia Provincial Museum 1. Victoria, BC: British Columbia Provincial Museum.

———. 1941. "Insularity in the Genus Sorex on the North Coast of British Columbia." *Proceedings of the Biological Society of Washington* 54: 95–108.

Moon, B.R., and W.P. Leonard. 2001. "A Semiarboreal Nest of the American Shrew-mole, *Neurotrichus gibbsii*." *Northwestern Naturalist* 82, no. 1: 26–27.

Moore, A.W. 1940. "A Live Mole Trap." *Journal of Mammalogy* 21, no. 2: 223–225.

Moore, J.W., and G.J. Kenagy. 2004. "Consumption of Shrews, *Sorex* spp., by Arctic Grayling, *Thymallus Arcticus*." *Canadian Field-Naturalist* 118, no. 1: 111–114.

Moya-Costa, R., G. Cuenca-Bescós, B. Bauluz and J. Rofes. 2018. "Structure and Composition of Tooth Enamel in Quaternary Soricines (Mammalia)." *Quaternary International* 481: 52–60.

Munro, J.A. 1947. *Observations of Birds and Mammals in Central British Columbia.* Occasional Papers of the British Columbia Provincial Museum 6. Victoria, BC: British Columbia Provincial Museum.

———. 1955. "Additional Observations of Birds and Mammals in the Vanderhoof Region, British Columbia." *American Midland Naturalist* 53, no. 1: 56–60.

Mycroft, E.E., A.B.A. Shafer and D.T. Stewart. 2011. "Cytochrome-*b* Sequence Variation in Water Shrews (*Sorex palustris*) from Eastern and Western North America." *Northeastern Naturalist* 18, no. 4: 497–508.

Nagorsen, D. 2016. *Small Mammal Management and Conservation in British Columbia: Assessment of Knowledge Gaps and Research Needs*. Wildlife Bulletin. Victoria, BC: BC Ministry of Environment and Climate Change Strategy.

Nagorsen, D.W., and D.M. Jones. 1981. "First Records of the Tundra Shrew (*Sorex tundrensis*) in British Columbia." *Canadian Field-Naturalist* 95, no. 1: 93–94.

Nagorsen, D.W., and N. Panter. 2009. "Identification and Status of the Olympic Shrew (*Sorex rohweri*) in British Columbia." *Northwestern Naturalist* 90, no. 2: 117–129.

Nagorsen, D.W., N. Panter and A.G. Hope. 2017. "Are the Western Water Shrew (*Sorex navigator*) and American Water Shrew (*Sorex palustris*) Morphologically Distinct?" *Canadian Journal of Zoology* 95, no. 10: 727–736.

Nagorsen, D.W., G.G.E. Scudder, D.J. Huggard, H. Stewart and N. Panter. 2001. "Merriam's Shrew, *Sorex merriami*, and Preble's Shrew, *Sorex preblei*: Two New Mammals for Canada." *Canadian Field-Naturalist* 115, no. 1: 1–8.

Newman, J.R. 1976. "Population Dynamics of the Wandering Shrew *Sorex vagrans*." *Wasmann Journal of Biology* 34, no. 2: 235–250.

Nováková, L., J. Lázaro, M. Muturi, C. Dullin and D.K.N. Dechmann. 2022. "Winter Conditions, Not Resource Availability Alone, May Drive Reversible Seasonal Skull Size Changes in Moles." *Royal Society Open Science* 9, no. 9. doi.org/10.1098/rsos.220652.

Nussbaum, R.A., and C. Maser. 1969. "Observations of *Sorex palustris* Preying on *Dicamptodon ensatus*." *Murrelet* 50, no. 2: 23–24.

Oaten, D.K., and K.W. Larsen. 2008. "Aspen Stands as Small Mammal 'Hotspots' within Dry Forest Ecosystems of British Columbia." *Northwest Science* 82, no. 4: 276–285.

O'Neill, M.B., D.W. Nagorsen and R.J. Baker. 2005. "Mitochondrial DNA Variation in Water Shrews (*Sorex palustris*, *Sorex bendirii*) from Western North America: Implications for Taxonomy and Phylogeography." *Canadian Journal of Zoology* 83, no. 11: 1469–1475.

Pattie, D.L. 1969. "Behavior of Captive Marsh Shrews (*Sorex bendirii*)." *Murrelet* 50, no. 3: 28–32.

Pattie, D. 1973. "Sorex bendirii." *Mammalian Species*, no. 27: 1–2.

Patton, J.L., and C.J. Conroy. 2017. "The Conundrum of Subspecies: Morphological Diversity among Desert Populations of the California Vole (*Microtus californicus*, Cricetidae)." *Journal of Mammalogy* 98, no. 4: 1010–1026.

Pedersen, R.J. 1963. "The Life History and Ecology of Townsend's Mole *Scapanus townsendii* (Bachman) in Tillamook County Oregon." MS thesis, Oregon State University.

———. 1966. "Nesting Behavior of Townsend's Mole." *Murrelet* 47, no. 2: 47–48.

Peltonen, A., S. Peltonen, P. Vilpas and A. Beloff. 1989. "Distributional Ecology of Shrews in Three Archipelagoes in Finland." *Annales Zoologici Fennici* 26, no. 4: 381–387.

Pernetta, J.C. 1977. "Anatomical and Behavioural Specialisations of Shrews in Relation to Their Diet." *Canadian Journal of Zoology* 55, no. 9: 1442–1453.

Pocock, M.J.O., and N. Jennings. 2006. "Use of Hair Tubes to Survey for Shrews: New Methods for Identification and Quantification of Abundance." *Mammal Review* 36, no. 4: 299–308.

Quay, W.B. 1951. "Observations on Mammals of the Seward Peninsula, Alaska." *Journal of Mammalogy* 32, no. 1: 88–99.

Racey, K. 1929. "Observations on *Neurotrichus gibbsii gibbsii*." *Murrelet* 10, no. 3: 61–62.

———. 1936. "Notes on Some Mammals of the Chilcotin, British Columbia." *Canadian Field-Naturalist* 50, no. 2: 15–21.

Rausch, R.L., J.E. Feagin and V.R. Rausch. 2007. "*Sorex rohweri* sp. nov. (Mammalia, Soricidae) from Northwestern North America." *Mammalian Biology* 72, no. 2: 93–105.

Reed, C.A. 1951. "Locomotion and Appendicular Anatomy in Three Soricoid Insectivores." *American Midland Naturalist* 45, no. 3: 513–671.

Rust, A.K. 1978. "Activity Rhythms in the Shrews, *Sorex sinuosus* Grinnell and *Sorex trowbridgii* Baird." *American Midland Naturalist* 99, no. 2: 369–382.

Rychlik, L., I. Ruczynski and Z. Borowski. 2010. "Radiotelemetry Applied to Field Studies of Shrews." *Journal of Wildlife Management* 74, no. 6: 1335–1342.

Ryckman, J. 2020. "Ecological and Genetic Connectivity of Shrews (*Sorex* spp.) across Interstate-90 in the Washington Cascade Range." MS thesis, Central Washington University.

Ryder, G.R. 2010. "Field Observation of Caravanning by a Family of Pacific Water Shrews in British Columbia." *Wildlife Afield* 7, no. 2: 298–300.

Ryder, G.R., and R.W. Campbell. 2007. "First Pacific Water Shrew Nest for British Columbia." *Wildlife Afield* 4, no. 1: 74–75.

Rzebik-Kowalska, B. 2003. "Distribution of Shrews (Insectivora, Mammalia) in Time and Space." *Deinsea* 10, no. 1: 499–508.

Salt, J.R. 2005. "Habitat Preferences of Arctic Shrew in Central and Southern Alberta." *Blue Jay* 63, no. 2: 85–86.

Schaefer, V.H. 1978. "Aspects of Habitat Selection in the Coast Mole (*Scapanus orarius*) in British Columbia." PhD diss., Simon Fraser University.

———. 1982. "Movements and Diel Activity of the Coast Mole *Scapanus orarius* True." *Canadian Journal of Zoology* 60, no. 3: 480–482.

Schaefer, V.H., and R.M.F. Sadleir. 1979. "Concentrations of Carbon Dioxide and Oxygen in Mole Tunnels." *Acta Theriologica* 24, no. 21: 267–276.

———. 1981. "Factors Influencing Molehill Construction by the Coast Mole (*Scapanus orarius* True)." *Mammalia* 45, no. 1: 31–38.

Schowalter, D.B. 2002. "Records of Pygmy Shrew, Northern Bog Lemming, and Heather Vole from Owl Pellets from North-Central Alberta." *Alberta Naturalist* 32, no. 1: 72–73.

Sealy, S.G. 2017. "Dead Shrews on the Road: Discarded by Mammalian Predators?" *Blue Jay* 75, no. 3: 18–22.

Seip, D.R., and J. Savard. 1991. *Maintaining Wildlife Diversity in Old Growth Forests and Managed Stands*. Annual Progress Report 1989–1990. Victoria, BC: British Columbia Ministry of Forests, Research Branch.

Sheehan, S.T., and C. Galindo-Leal. 1996. "Townsend's Mole (*Scapanus townsendii*) in the Lower Fraser Valley: Distribution, Habitat, Densities and Habitat Management." Unpublished report. Centre for Applied Conservation Biology, University of British Columbia, Vancouver.

———. 1997. "Identifying Coast Moles, *Scapanus orarius*, and Townsend's Mole, *Scapanus townsendii*, from Tunnel and Mound Size." *Canadian Field-Naturalist* 111, no. 3: 463–465.

Siemers, B.M., G. Schauermann, H. Turni and S. von Merten. 2009. "Why Do Shrews Twitter? Communication or Simple Echo-Based Orientation." *Biology Letters* 5, no. 5: 593–596.

Smith, M.E., and M.C. Belk. 1996. "Sorex monticolus." *Mammalian Species*, no. 528: 1–5.

Sorenson, M.W. 1962. "Some Aspects of Water Shrew Behavior." *American Midland Naturalist* 68, no. 2: 445–462.

Stewart, D.T., T.B. Herman and T. Teferi. 1989. "Littoral Feeding in a High-Density Insular Population of *Sorex cinereus*." *Canadian Journal of Zoology* 67, no. 8: 2074–2077.

Stewart, D.T., M. McPherson, J. Robichaud and L. Fumagalli. 2003. "Are There Two Species of Pygmy Shrews (*Sorex*)? Revisiting the Question Using DNA Sequence Data." *Canadian Field-Naturalist* 117, no. 1: 82–88.

Stewart, D.T., N. Perry and L. Fumagilli. 2002. "The Maritime Shrew, *Sorex maritimensis* (Insectivora: Soricidae): A Newly Recognized Canadian Endemic." *Canadian Journal of Zoology* 80, no. 1: 94–99.

Stinson, D.W., and J.D. Reichel. 1985. "Rediscovery of the Pygmy Shrew in Washington." *Murrelet* 66, no. 2: 59–60.

Stromgren, E.J., and T.P. Sullivan. 2014. "Influence of Pitfall versus Longworth Livetraps, Bait Addition, and Drift Fences on Trap Success and Mortality of Shrews." *Acta Theriologica* 59, no. 1: 203–210.

Sullivan, D.S., and T.P. Sullivan. 1982. "Effects of Logging Practices and Douglas-fir, *Pseudotsuga menziesii*, Seeding on Shrew, *Sorex* spp., Populations in Coastal Coniferous Forest in British Columbia." *Canadian Field-Naturalist* 96: 455–461.

Svihla, A. 1934. "The Mountain Water Shrew." *Murrelet* 15, no. 2: 44–45.

Teferi, T., T.B. Herman and D.T. Stewart. 1992. "Breeding Biology of an Insular Population of the Masked Shrew (*Sorex cinereus* Kerr) in Nova Scotia." *Canadian Journal of Zoology* 70, no. 1: 62–66.

Terry, C.J. 1978. "Food Habits of Three Sympatric Species of Insectivora in Western Washington." *Canadian Field-Naturalist* 92, no. 1: 38–44.

———. 1981. "Habitat Differentiation among Three Species of Sorex and Neurotrichus gibbsii in Washington." *American Midland Naturalist* 106, no. 1: 119–125.

Tye, S.P., K. Geluso and M.R. Fugagli. 2016. *Merriam's Shrew (*Sorex merriami*) in the Diet of a Mexican Spotted Owl (*Strix occidentalis lucida*) from Grant County, New Mexico.* Occasional Papers 341. Lubbock, TX: Museum of Texas Tech University.

van Tighem, K.J., and L.W. Gyug. 1984. *Ecological Land Classification of Mount Revelstoke and Glacier National Parks, British Columbia*. Vol. 2, *Wildlife Resource*. Calgary, AB: Canadian Wildlife Service.

van Zyll de Jong, C.G. 1980. "Systematic Relationships of Woodland and Prairie Forms of the Common Shrew, *Sorex cinereus cinereus* Kerr and *S. c. haydeni* Baird, in the Canadian Prairie Provinces." *Journal of Mammalogy* 61, no. 1: 66–75.

———. 1983a. *Handbook of Canadian Mammals*. Vol. 1 of *Marsupials and Insectivores*. Ottawa, ON: National Museums of Canada, National Museum of Natural Sciences.

———. 1983b. "A Morphometric Analysis of North American Shrews of the *Sorex arcticus* Group, with Special Consideration of the Taxonomic Status of *S. a. maritimensis*." *Le Naturaliste Canadien* 110, no. 4: 373–378.

Verts, V.J., and L.N. Carraway. 1998. *Land Mammals of Oregon*. Berkeley, CA: University of California Press.

Whitaker, J.O., Jr. 2004. "Sorex cinereus." *Mammalian Species*, no. 743: 1–9.

Whitaker, J.O., Jr., S.P. Cross and C. Maser. 1983. "Food of Vagrant Shrews (*Sorex vagrans*) from Grant County, Oregon, as Related to Livestock Grazing Pressures." *Northwest Science* 57, no. 2: 107–111.

Whitaker, J.O., Jr., and C. Maser. 1976. "Food Habits of Five Western Oregon Shrews." *Northwest Science* 50, no. 2: 102–107.

Whitaker, J.O., Jr., C. Maser and R.J. Pedersen. 1979. "Food and Ectoparasitic Mites of Oregon Moles." *Northwest Science* 53, no. 4: 268–273.

Whitaker, J.O., Jr., and L.L. Schmeltz. 1973. "Food and External Parasites of *Sorex palustris* and Food of *Sorex cinereus* from St. Louis County, Minnesota." *Journal of Mammalogy* 54, no. 1: 283–285.

Woodman, N. 2018. *American Recent Eulipotyphla: Nesophontids, Solenodons, Moles, and Shrews in the New World*. Washington, DC: Smithsonian Institution Scholarly Press.

Woodman, N., and R.D. Fisher. 2016. "Identification and Distribution of the Olympic Shrew (Eulipotyphla: Soricidae), *Sorex rohweri* Rausch et al., 2007 in Oregon and Washington, Based on USNM Specimens." *Proceedings of the Biological Society of Washington* 129, no. 1: 84–102.

Wrigley, R.E., J.E. Dubois and H.W.R. Copland. 1979. "Habitat, Abundance, and Distribution of Six Species of Shrews in Manitoba." *Journal of Mammalogy* 60, no. 3: 505–520.

Yates, T.L. 1978. "The Systematics and Evolution of North American Moles (Insectivora: Talpidae)." PhD diss., Texas Tech University.

Youngman, P.M. 1975. *Mammals of the Yukon Territory*. Publications in Zoology 10. Ottawa, ON: National Museum of Natural Sciences, National Museums of Canada.

Zuleta, G.A., and C. Galindo-Leal. 1994. *Distribution and Abundance of Four Species of Small Mammals at Risk in a Fragmented Landscape*. No. WR-64. Victoria, BC: BC Ministry of Environment, Lands and Parks, Wildlife Branch.

Credits

Cover and interior design by Jeff Werner
Illustrations by Michael Hames, except pages 123, 159 and 165 by Donald Gunn
Maps on pages 73, 81, 89, 97, 103, 111, 119, 125, 131, 139, 145, 153, 161, 167, 173, 179 and 185 by BC Conservation Data Centre
Maps on pages 7 and 13 by Royal BC Museum
Map on page 69 by Jeff Werner
Editing by Eva van Emden
Copy editing by Grace Yaginuma
Proofreading by Eva van Emden
Index by François Trahan

Front cover photograph of AMERICAN WATER SHREW, *Sorex palustris*, pursuing a crayfish by Kenneth Catania
Back cover photograph of TOWNSEND'S MOLE, *Scapanus townsendii*, by Andrew Hendry

John Acorn: 23 left two, 106.
Bob Brett: 71.
Rob Cannings: 11 bottom right two.
Syd Cannings: 9 top, 180, 194.
Darren Copley: 23 top right.
Vanessa Craig: 46.
Michael Hames: 49.
Andrew Hendry: 87.
Jared Hobbs: 105.
Andrew Hope: 109.
Bengul Kurtar: 8 bottom, 11 top.
Markus Merkens: 30.
Daphne Nagorsen: 8 top, 11 bottom left, 132.
David Nagorsen: 3, 9 bottom, 21, 23 right bottom three, 31 bottom, 44 top right & bottom two, 50, 51, 56 second & fourth, 72 top two, 80 top two & far right, 82, 88 top two & far right, 90, 96 top three, 102 top three, 110 top three, 118 top three, 124 top three, 130 top four, 133, 138 top three, 144 top three, 152 top three, 155, 160 top three, 162, 163, 166 top three, 169, 172 top three, 178 top three, 184 top three, 186, 187, 192, 201 bottom two, 239 bottom.
Nancy Nagorsen: 239 top.

Nick Panter: 17, 19, 31 top, 48 bottom, 53, 54, 55, 56 first & third, 57, 58, 59, 60, 61, 62, 63, 64, 65, 72 bottom three, 80 left bottom three, 88 far left bottom three, 96 bottom four, 102 bottom four, 110 bottom four, 118 bottom four, 124 bottom four, 130 bottom four, 138 bottom four, 144 bottom four, 152 bottom four, 160 bottom four, 166 bottom four, 172 bottom four, 178 bottom four, 184 bottom four, 201 top.

Caroline Penn: 147.

Jordan Ryckman: 44 top left, 171.

Michael Schmidt Photography Vancouver: 48 top, 72 middle two.

Andy Teucher: 126.

Douglas Watkinson: 156.

David Wong: 79.

Index

Figures and illustrations indicated by page numbers in **bold**.

About the Authors

David Nagorsen is a research associate at the Royal BC Museum (Victoria) and the Royal Ontario Museum (Toronto). He has more than 30 years' experience as a mammalogist and museum biologist carrying out research, fieldwork, endangered species conservation and public education. Following his museum career, he worked as a wildlife consultant for the British Columbia, Yukon and federal governments as well as various consulting companies. His fascination with shrews began in the 1970s while working on a small-mammal population study. He has authored or co-authored five handbooks on BC mammals and many scientific papers and reports.

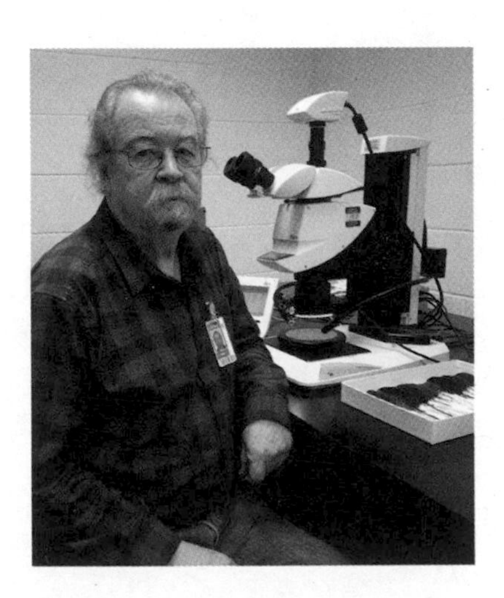

Nick Panter is a retired biologist and a volunteer at the Royal BC Museum. He has worked as a museum preparator and collection manager for vertebrate collections at the University of Alberta and the Royal BC Museum. He has had a long-term interest in small mammals, particularly shrews and chipmunks, and has collaborated on several scientific papers about them.